IB Mathematics:

Applications and Interpretation SL

in **70** pages

2021 Edition

George Feretzakis, B.Sc., M.Sc., Ph.D.

This book has been developed independently of the International Baccalaureate Organization (IBO), and the content is in no way connected with nor endorsed by the International Baccalaureate Organization (IBO).

While every attempt has been made to trace and acknowledge copyright, the author apologizes for any accidental infringement where the copyright has proved untraceable.

Cover Image: iStock.com/Tramont_ana

Ed. (Mar. 2021)

The best IB Math revision guide app, with video lessons, many calculators, solvers, and it is **100% free**, No in-app purchases, No advertisements.

Any comments, improvements, or suggestions can be sent directly to the author at george@mathematics4u.com.

CONTENTS

Introduction

This revision guide will be a valuable resource and reference for students, assisting them in understanding and learning the theory of IB Mathematics: Applications and Interpretation Standard Level.

The ideal preparation for any student preparing for the IB math exams is to systematically practice doing questions from past papers. Similar exercises can be found in any IB Mathematics: Applications and Interpretation SL textbook and on official distributor sites for IB materials.

The guide aims to help the IB student by revising the theory and going through some well-chosen examples of the new IB Mathematics: Applications and Interpretation SL curriculum.

I have made a concerted effort to explain the mathematical terms to the student clearly, straightforward, and understandable. I aim to create **thorough** and **concise** material with an emphasis on **simplicity,** which will be effective for both teachers and students.

By presenting the theory that every IB student should know before taking any quiz, test, or exam, this Revision Guide is designed to make the topics of IB Math Applications and Interpretation SL both comprehensible and easy to grasp.

Dr. George Feretzakis, November 2020

"Truth is ever to be found in the simplicity, and not in the multiplicity and confusion of things."

Isaac Newton

Ⓐ Get the app

The best IB Math revision guide app, with video lessons, many calculators, solvers, and it is **100% free**, **No** in-app purchases, **No** advertisements.

Useful Formulas

$(\alpha \pm b)^2 = \alpha^2 \pm 2\alpha b + b^2$		$\alpha^2 - b^2 = (\alpha + b)(\alpha - b)$	
Area of a parallelogram b: base, h:height	$A = b \times h$	Volume of a cylinder	$V = \pi r^2 h$
Area of a triangle	$A = \frac{1}{2}(b \times h)$	Volume of a cuboid	$V = l \times w \times h$
Area of a trapezium	$A = \frac{1}{2}(a + b) \times h$	Area of the curved surface of a cylinder	$A = 2\pi r h$
Area of a circle	$A = \pi r^2$	Volume of a sphere	$V = \frac{4}{3}\pi r^3$
Circumference of a circle	$C = 2\pi r$	Volume of a cone	$V = \frac{1}{3}\pi r^2 h$

Surface area of a sphere	$A = 4\pi r^2$	Area of the curved surface of a cone	$A = \pi r l$
Volume of a pyramid	$V = \frac{1}{3}(Area) \times h$	Volume of a prism	$V = (Area) \times h$

Laws of Exponents

If a, b are positive numbers, x, y are real numbers and m, n are positive integers then:

$a^{x+y} = a^x a^y$	$\left(\frac{a}{b}\right)^x = \frac{a^x}{b^x}$	$a^{\frac{m}{n}} = \sqrt[n]{a^m}$
$a^{x-y} = \frac{a^x}{a^y}$	$a^0 = 1$	$a^{\frac{1}{n}} = \sqrt[n]{a}$
$(a^x)^y = a^{xy}$	$a^{-x} = \frac{1}{a^x}$	$\left(\frac{a}{b}\right)^{-x} = \left(\frac{b}{a}\right)^x$
$(ab)^x = a^x b^x$	$a^x = \frac{1}{a^{-x}}$	$\frac{a^{-x}}{b^{-y}} = \frac{b^y}{a^x}$

Logarithms

$log_a x = y \Leftrightarrow x = a^y$ where $a, x > 0, a \neq 1$	$log_a 1 = 0$	$log_a a = 1$	$x = a^{log_a x}$
$log x = y \Leftrightarrow x = 10^y$	$log\ 1 = 0$	$log\ 10 = 1$	$x = 10^{log x}$
$ln x = y \Leftrightarrow x = e^y$	$ln\ 1 = 0$	$ln\ e = 1$	$x = e^{ln x}$

Number and Measurements

Number Sets

- The set of **natural** numbers $\mathbb{N}$ is $\{0,1,2,3,\dots\}$.

- The set of **integers** $\mathbb{Z}$ is $\{-3,-2,-1,0,1,2,3,\dots\}$.

- The set of **rational** numbers $\mathbb{Q}$ is $\left\{\frac{a}{b}, b \neq 0\right\}$, where a,b are integers.

- **Irrational** numbers are all real numbers that cannot be expressed as a ratio of integers.

  For example: $\sqrt{7}$, $\sqrt{3}$, π

- The set of positive integers $\mathbb{Z}^+$ is $\{1,2,3,\dots\}$.

- The set of positive rational numbers is denoted by $\mathbb{Q}^+$.

- The set of positive real numbers is denoted by $\mathbb{R}^+$.

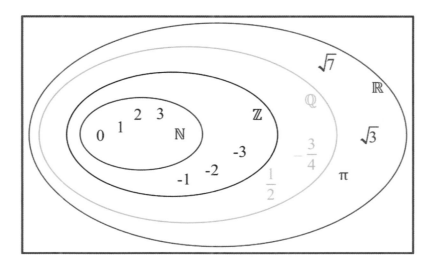

International System (SI) units

Base Quantity	Name	Symbol	Base Quantity	Name	Symbol
Length	meter	m	Area	square meter	m^2
Mass	kilogram	kg	Volume	cubic meter	m^3
Time	second	s	Velocity	meters per second	$m\,s^{-1}$
Temperature	kelvin	K	Density	kilogram per cubic meter	$kg\,m^{-3}$

Rules of Rounding

If the number you are rounding is followed by 5, 6, 7, 8, or 9, round the number up.

If the number you are rounding is followed by 0, 1, 2, 3, or 4, round the number down.

Numbers can be rounded to the nearest ten, the nearest hundred, the nearest thousand, and so on.
For example,
4,738 rounded to the nearest ten is 4,740
4,738 rounded to the nearest hundred is 4,700
4,738 rounded to the nearest thousand is 5,000

Rules of Rounding for decimals

Rounding decimals works precisely the same way as rounding whole numbers. The only difference is that instead of rounding to tens, hundreds, thousands, and so on, you round to tenths, hundredths, thousandths, and so on.

5.7287 rounded to the nearest tenth is 5.7

2.1732 rounded to the nearest hundredth is 2.17

4.7873 rounded to the nearest thousandth is 4.787

Significant Figure Rules[*]

There are three rules for determining how many significant figures are in a number:

1. Non-zero digits are always significant.

2. Any zeros between two significant digits are significant.

3. A final zero or trailing zeros in the decimal portion only are significant.

Trailing zeros are a sequence of 0s in the decimal representation of a number, after which no other digits follow.

Example 1	Example 2
1254.04 rounded to 4 s.f. is 1254	0.030503062 rounded to 4 s.f. is 0.03050
1254.04 rounded to 3 s.f. is 1250	0.030503062 rounded to 3 s.f. is 0.0305
1254.04 rounded to 2 s.f. is 1300	0.030503062 rounded to 2 s.f. is 0.031
1254.04 rounded to 1 s.f. is 1000	0.030503062 rounded to 1 s.f. is 0.03

[*] You can download our app using the QR code on page ii, where there is a calculator that can round any number to as many significant figures as desired using proper rounding rules. Apart from the calculator mentioned above, our app has many other solvers and calculators accompanied by the corresponding theory.

Error and Percentage error

The **Error** can be calculated by subtracting the **Exact** value (v_E) from the **Approximate** value (v_A), ignoring any minus sign.

$$Error = v_A - v_E$$

The **Percentage Error** can be calculated from the error by dividing it by the Exact value (v_E).

$$Percentage\ Error = \left| \frac{v_A - v_E}{v_E} \right| \times 100\%$$

Example

Harry measured his height and found 5.9 feet. However, his actual height is 5.7 feet.

What is the percent error Harry made when he measured his height?

Solution

The percent error is given by the following formula:

$$Percentage\ Error = \left| \frac{5.9 - 5.7}{5.7} \right| \times 100\% = 3.509\ \%\ (4\ s.f.)$$

Scientific notation

Scientific notation (Standard form) is a convenient way of writing very large or very small numbers.

A number is written in scientific notation $a \times 10^m$ where $1 \leq a < 10$ and the exponent m is an integer.

Examples

$$784{,}345 = 7.84345 \times 10^5$$

$$0.0045 = 4.5 \times 10^{-3}$$

Note: Calculator notation of the form aEb is not acceptable for the exams.

For example, you have to write the output of the GDC $4.24E - 5$ as 4.24×10^{-5}

Upper and Lower bounds of rounded numbers

The **lower bound** is the smallest value that would round up to the estimated value.

The **upper bound** is the smallest value that would round up to the next estimated value.

For example, if you are given $24.1\ cm$ to the nearest $0.1\ cm$, then the upper bound is $24.15\ cm$, and the **lower bound** is $24.05\ cm$.

Sequences, Series and Compound Interest

A **sequence** is a set of numbers arranged in a definite order
$$u_1, u_2, u_3, \ldots, u_n, \ldots$$
, where $u_1, u_2, u_3, \ldots, u_n, \ldots$ are called **terms**.

Arithmetic Sequence and Series

A sequence $\{u_n\}$ in which each term differs from the previous one by the same constant d (common difference) is called an **arithmetic sequence**.
$$u_{n+1} - u_n = d$$
The general term u_n can be found using the formula
$$u_n = u_1 + (n-1)d$$

The Sum (S_n) of the first n terms of an arithmetic sequence $\{u_n\}$ is called an **arithmetic series** and given by

$$S_n = \frac{n}{2}(u_1 + u_n)$$

$$S_n = \frac{n}{2}[2u_1 + (n-1)d]$$

Example

An arithmetic sequence has a first term of 80 and the 24th term of 172. Find the 74th term, an expression for the general term, and the sum S_n.

Solution

Using the formula $u_n = u_1 + (n-1)d$, we have that
$$u_{24} = u_1 + (24-1)d \Rightarrow d = \frac{172 - 80}{23} = \frac{92}{23} = 4$$

Thus, $u_{74} = u_1 + (74-1)d = 80 + 73 \times 4 = 372$

The general term is $u_n = 80 + (n-1)4$

and the sum of the first n terms is $S_n = \frac{n}{2}[160 + (n-1)4]$

Geometric Sequence and Series

A sequence $\{u_n\}$ in which each term can be obtained from the previous one by multiplying a non-zero constant r (common ratio) is called **geometric sequence**.

$$\frac{u_{n+1}}{u_n} = r$$

The general term u_n can be found using the formula

$$u_n = u_1 r^{n-1}$$

The Sum (S_n) of the first n terms of a geometric sequence $\{u_n\}$ is called **geometric series** and given by

$$S_n = \frac{u_1(r^n - 1)}{r - 1} = \frac{u_1(1 - r^n)}{1 - r}, r \neq 1$$

Note: If we want to deduce the general term u_n of an **arithmetic** or a **geometric** sequence from S_n, we can use the following formula:

$$u_n = S_n - S_{n-1} \text{ and } u_1 = S_1$$

Example

A geometric sequence has a 4th term of 32 and a 7th term of 256. Find the 10th term, an expression for the general term u_n and the sum S_n.

Solution

Using the formula $u_n = u_1 r^{n-1}$ we have that
$$u_4 = u_1 r^{4-1} \Rightarrow 32 = u_1 r^3$$
$$u_7 = u_1 r^{7-1} \Rightarrow 256 = u_1 r^6$$

Dividing the two equations, we have $\dfrac{u_1 r^3}{u_1 r^6} = \dfrac{32}{256} \Leftrightarrow r^3 = 8 \Rightarrow r = 2$

and $32 = u_1 r^3 \Rightarrow 32 = u_1 2^3 \Rightarrow u_1 = 4$

Thus, $u_{10} = u_1 r^{n-1} = 4 \times 2^{10-1} = 2048$

The general term is $u_n = 4 \times 2^{n-1}$ and the sum $S_n = \frac{4(2^n - 1)}{2 - 1} = 4(2^n - 1)$

Sigma Notation

Sigma notation is a way of expressing sums uses the Greek letter Σ

$$\sum_{i=1}^{n} a_i = a_1 + a_2 + \cdots + a_n$$

, where i is the index of summation taking values from 1 to n.

Properties of Sigma Notation

$$\sum_{i=1}^{n}(a_i \pm b_i) = \sum_{i=1}^{n}a_i \pm \sum_{i=1}^{n}b_i$$	$$\sum_{i=1}^{n}ca_i = c\sum_{i=1}^{n}a_i$$
$$\sum_{i=1}^{n}c = cn$$	$$\sum_{i=k}^{n}a_i = \sum_{i=1}^{n}a_i - \sum_{i=1}^{k-1}a_i$$

Simple Interest

The simple interest formula for calculating the **Future Value (FV)** of an amount with a **Present Value (PV)**, where $r\%$ is the nominal annual interest rate, and n is the time in years:

$$FV = PV\left(1 + \frac{r}{100}n\right)$$

Compound Interest

The compound interest formula for calculating the **Future Value(FV)** of an amount with a **Present Value(PV)** is:

$$FV = PV\left(1 + \frac{r}{100k}\right)^{kn}$$

where $r\%$ is the nominal annual interest rate, n the **number of years** and k is the **number of compounding periods per year**. Compound interest can be calculated annually ($k = 1$), semiannually ($k = 2$), quarterly ($k = 4$), or monthly ($k = 12$).

Example

Marina invests \$80,000 in a savings account, earning 5% per year compounded monthly.

Calculate the value of the investment after 4 whole years.

Answer

In this example, $r = 5, n = 4,$ and $k = 12$

$$FV = PV\left(1 + \frac{r}{100k}\right)^{kn} = \$80,000\left(1 + \frac{5}{100 \times 12}\right)^{12\times 4} =$$

$$= \$97,671.63 = \$97,700 \; (3 \; s.f.)$$

Annual Depreciation

Depreciation is defined as the reduction of the value of a business asset until the value of the asset becomes zero or negligible.

The depreciation formula for calculating the **Future Value(FV)** of an asset with a **Present Value(PV)** is:

$$FV = PV \left(1 + \frac{r}{100}\right)^n$$

where $r\%$ is the **depreciation rate per period** (a negative number) and n the **number of periods**.

Loans

For the loan repayment period, borrowers will have a fixed repayment schedule. In regular installments, payments will be made in a fixed amount consisting of principal and interest. Student loans, car loans, and home mortgages are typical examples of amortized loans.

Amortization is the process of spreading out a loan into a series of fixed payments over time.

Optional: The formula for calculating the **Payment Amount per Period (PMT)** is:

$$PMT = \frac{P\,r}{1 - (1 + r)^{-n}}$$

where r is the **interest rate per period** expressed as a decimal, P is the **principal** (loan amount), and n the **number of periods**.

Example

Gregory receives a loan of $24,000 to buy a car at an annual interest rate of 8% compounded monthly. The loan is taken for 4 years and will be paid in monthly repayments. Calculate the monthly installments and the total interest charged.

Answer

TI 84 +	Casio fx9860 series, fx-CG20, fx-CG50
Press **APPS** to select the Finance application **1: Finance** ▶ **1: TVM solver** and then enter the known values for the TVM variables: $n = 4 \times 12 = 48, I\% = 8,\ PV = -24,000, FV = 0, P/Y = 12,$ $$C/Y = 12, PMT: END$$ Place the cursor on the TVM variable for which you want to solve (PMT) and press **alpha** **enter (solve)** $$PMT = 585.91$$	Press **MENU** **Financial** **EXE** then select **F4: Amortization** and fill out the required information: $n = 4 \times 12 = 48,$ $$I\% = 8, PV = -24,000,$$ $$FV = 0, P/Y = 12, C/Y = 12$$ ▶ **CMPD** ▶ **PMT** ▶ $PMT = 585.91$

So, the monthly installment (repayments) for 4 years (48 months) is $585.91, and the total amount that will be paid is $585.91 \times 48 = \$28,123.68$.

Therefore, the total interest charged is $\$28,123.68 - \$24,000.00 = \$4,123.68$.

Straight Lines

The equation of a straight line is usually written:

$$y = mx + c$$

where m is the **slope** or gradient, and c is the **y-intercept**.

The **standard** or **general** form of a straight line is

$$ax + by + c = 0$$

Another way to find the equation of a straight line is the following:

$$y - y_1 = m(x - x_1)$$

, where m is the slope and (x_1, y_1) is a point that lies on the line.

To find the **slope m**, you use the following formula:

$$m = \frac{y_2 - y_1}{x_2 - x_1}$$

where (x_1, y_1) and (x_2, y_2) are two points that lie on the line.

Note: When a line has a **positive** slope, it rises left to right (its graph will be **increasing**)

When a line has a **negative** slope, it falls left to right (its graph will be **decreasing**)

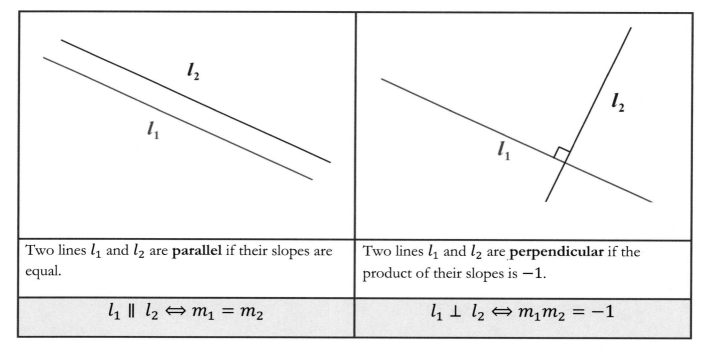

Two lines l_1 and l_2 are **parallel** if their slopes are equal.	Two lines l_1 and l_2 are **perpendicular** if the product of their slopes is -1.
$l_1 \parallel l_2 \Leftrightarrow m_1 = m_2$	$l_1 \perp l_2 \Leftrightarrow m_1 m_2 = -1$

Midpoint Formula: The coordinates of the midpoint $M(x_M, y_M)$ of two points $A(x_1, y_1)$ and $B(x_2, y_2)$, are given by the formulas: $$x_M = \frac{x_1 + x_2}{2}, \qquad y_M = \frac{y_1 + y_2}{2}$$ In 3-D: The coordinates of the $M(x_M, y_M, z_M)$ of two points $A(x_1, y_1, z_1)$ and $B(x_2, y_2, z_2)$ are: $$x_M = \frac{x_1 + x_2}{2}, \qquad y_M = \frac{y_1 + y_2}{2}, \qquad z_M = \frac{z_1 + z_2}{2}$$	**Distance Formula**: Given two points $A(x_1, y_1)$ and $B(x_2, y_2)$, the distance (d) between these two points is given by the formula: $$d = \sqrt{(x_1 - x_2)^2 + (y_1 - y_2)^2}$$ In 3-D: Given two points $A(x_1, y_1, z_1)$ and $B(x_2, y_2, z_2)$, the distance (d) between these two points is given by the formula: $$d = \sqrt{(x_1 - x_2)^2 + (y_1 - y_2)^2 + (z_1 - z_2)^2}$$
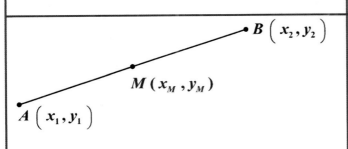	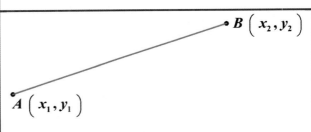

Example

Find the equation of the **perpendicular bisector** of the line segment AB, where $A(2, -3)$ and $B(4,3)$.

Answer

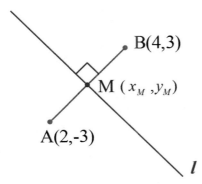

The coordinates of the midpoint M are given by the formulas

$$x_M = \frac{x_A + x_B}{2} = \frac{2 + 4}{2} = 3$$

$$y_M = \frac{y_A + y_B}{2} = \frac{3 + (-3)}{2} = 0$$

So, $M(3,0)$

The gradient of the line segment AB is $m = \frac{y_B - y_A}{x_B - x_A} = \frac{3 - (-3)}{4 - 2} = \frac{6}{2} = 3$

The line l (the perpendicular bisector) is perpendicular to line segment AB; therefore its slope is the negative reciprocal of 3. Now, both the point $M(3,0)$, which lies on the line l and the slope $\left(-\frac{1}{3}\right)$ of the line are known. Therefore, the equation of the straight line l is given by the following formula:

$$y - 0 = -\frac{1}{3}(x - 3) \text{ or } y = -\frac{1}{3}x + 1$$

Quadratic Equations & Functions

To **solve a quadratic equation** of the form $ax^2 + bx + c = 0, a \neq 0$, follow these steps:

1. When the discriminant $(\Delta = b^2 - 4ac)$ is **positive** $(\Delta > 0)$ then the equation has two distinct real roots r_1 and r_2.

$$r_{1,2} = \frac{-b \pm \sqrt{b^2 - 4ac}}{2a}$$

2. When the discriminant is equal to **zero** $(\Delta = 0)$ then the equation has one double (two equal real roots) root r.

$$r = \frac{-b}{2a}$$

3. When the discriminant is **negative** $(\Delta < 0)$ then the equation has no real roots.

A quadratic function is one of the form

$$f(x) = ax^2 + bx + c, \text{ where } a, b, c \in \mathbb{R} \text{ and } a \neq 0.$$

The graph of a quadratic function is a curve called a parabola. Parabolas may open upward or downward, and all have the same basic "U" shape.

If a is **positive**, the graph **opens upward** (Figure 1), and if a is **negative** (Figure 2), then it **opens downward**. The **y-intercept** of the above quadratic function is c.

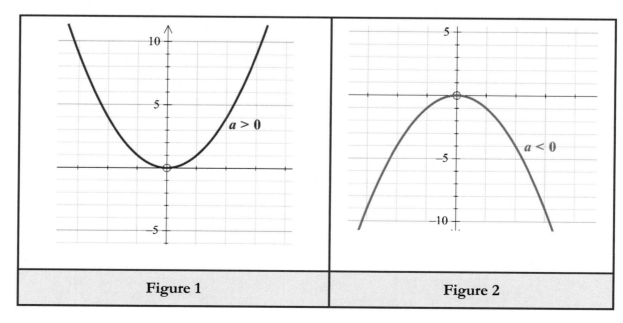

| Figure 1 | Figure 2 |

All parabolas are symmetric with respect to a line called the **axis of symmetry**, with the equation:

$$x = \frac{-b}{2a}$$

A parabola intersects its axis of symmetry at a point called the vertex **V** of the parabola, which has coordinates:

$$V\left(\frac{-b}{2a}, f\left(\frac{-b}{2a}\right)\right)$$

The three most common forms that are used to express quadratic functions are:

Standard form: $f(x) = ax^2 + bx + c$

Factored form: $f(x) = a(x - r_1)(x - r_2)$, where r_1, r_2 are the two distinct real roots which are the x-intercepts (zeros) of the graph $f(x)$. The equation of the axis of symmetry (i.e., the x-coordinate of the vertex) passes through the mid-point of the x-intercepts of the parabola.

$f(x) = a(x - r)^2$, where r is a real double root (two equal real roots) which is the x-intercept of the graph of $f(x)$.

Vertex form: $f(x) = a(x - h)^2 + k$, where h, k are the coordinates of the vertex $V(h, k)$.

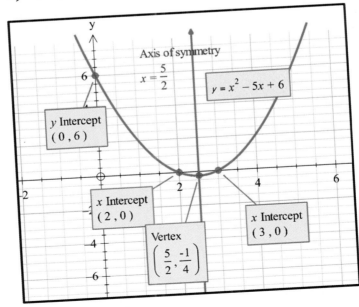

For example, the quadratic function $f(x) = x^2 - 5x + 6$ can be written as

Standard form: $f(x) = x^2 - 5x + 6$

Factored form: $f(x) = 1(x - 2)(x - 3)$, where $2, 3$ are the two x-intercepts (zeros).

Vertex form: $f(x) = 1(x - \frac{5}{2})^2 - \frac{1}{4}$, where the vertex V has coordinates $(\frac{5}{2}, -\frac{1}{4})$.

Note: We observe that the x-coordinate of the vertex is the midpoint of the two x-intercepts, $\frac{2+3}{2} = \frac{5}{2}$.

Example

Find the equation of the quadratic function, which passes through the point $(1,2)$, and its graph cuts the x-axis at $2, 3$.

Answer

The quadratic equation can be written in the form $f(x) = a(x - r_1)(x - r_2)$, where r_1, r_2 are the two x-intercepts.

So, the equation is $f(x) = a(x - 2)(x - 3)$, where a a non-zero constant.

To find the constant a, we plug in the coordinates of the point $(1,2)$.

$$2 = a(1 - 2)(1 - 3)$$
$$2 = a(-1)(-2)$$
$$2 = 2a$$
$$a = 1$$

Therefore, the equation of the quadratic function is

$$f(x) = 1(x - 2)(x - 3) = x^2 - 5x + 6$$

Functions

Relation: A relation is any set of ordered-pair numbers.

For example, let the relation $R = \{(1, 15), (2, 17), (3, 18), (4, 26), (4, 67)\}$

The set of all first elements is called the **domain** of the relation.
   The domain of R is the set $\{1, 2, 3, 4\}$.

The set of second elements is called the **range** of the relation.
   The range of R is the set $\{15, 17, 18, 26, 67\}$.

Function: A function is a relation in which **no** two ordered pairs have the same first element.

A function associates each element in its domain with **one and only one** element in its range.

All functions are relations, but not all relations are functions.

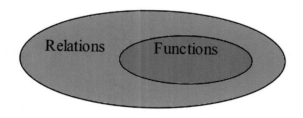

Example

Determine whether the following relations are functions

a) $A = \{(0, 5), (3, 7), (6, 9), (8, 15)\}$
b) $B = \{(10, 13), (10, 31), (20, 15), (41, 23)\}$

Answer

a) $A = \{(0, 5), (3, 7), (6, 9), (8, 15)\}$ is a function because all the first elements are different.

b) $B = \{(10, 13), (10, 31), (20, 15), (41, 23)\}$ is not a function because the first element, **10**, is repeated.

The **domain (D_f) of a function f** is the set of all allowable values of the independent variable, commonly known as the x-values.

1. You cannot have a negative number under a square root or any even radical.
2. You cannot have zero in the denominator.
3. You cannot have a negative number or zero as an argument of a logarithmic function.
4. You cannot have a negative number or zero or one as the base of a logarithmic function.

The **range (R_f)** of a **function f** is the set of y-values when all x-values in the domain are evaluated into the function.

Vertical Line Test

The vertical line test is a method to determine if a relation is a function. A relation is a function if any vertical line intersects the graph in at most one point.

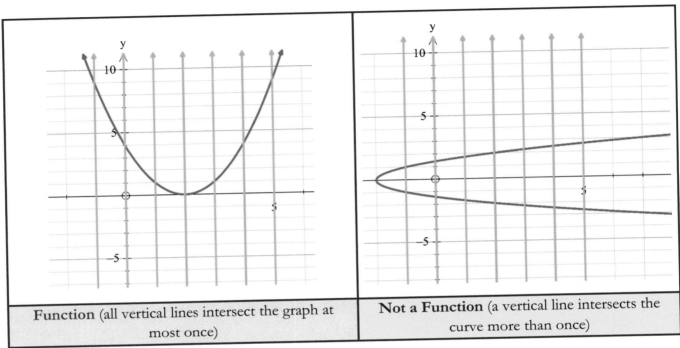

| Function (all vertical lines intersect the graph at most once) | Not a Function (a vertical line intersects the curve more than once) |

Inverse Function

The inverse of the function f is denoted by f^{-1} and is pronounced "f inverse" and <u>it's not</u> the reciprocal of f $\left(f^{-1}(x) \neq \frac{1}{f(x)}\right)$.

To determine algebraically the formula for the inverse of a function $y = f(x)$, you switch y and x to get $x = f(y)$ and then solve for y to get $y = f^{-1}(x)$.

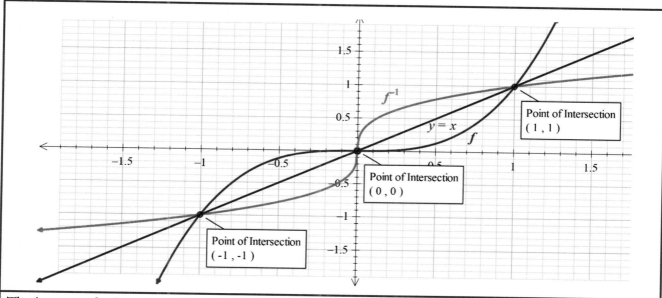

The inverse of a function differs from the function in that all the x-coordinates and y-coordinates have been switched. That is, if, for example, $(4,9)$ is a point on the graph of the function, then the point $(9,4)$ lies on the graph of the inverse function.

The graph of a function and its inverse are mirror images of each other. They are reflected in the identity function $y = x$.

Important: If the graphs of a function and its inverse intersect at one point, then this point will be on the line $y = x$, as shown in the figure above. Therefore, if we want to find the point(s) of intersection between $f(x)$ and $f^{-1}(x)$, instead of finding $f^{-1}(x)$ and then equating both of the functions, we could set $f(x) = x$ and find the common points between $f(x)$ and $y = x$.

Note: A function is said to be a **self-inverse** if $f(x) = f^{-1}(x)$ for all x in the domain.

For example, the reciprocal function $f(x) = \frac{1}{x}, (x \neq 0)$ is self-inverse.

Notes: The domain of f^{-1} is equal to the range of f.

The range of f^{-1} is equal to the domain of f.

For example, if $f(4) = 9$ then $f^{-1}(9) = 4$

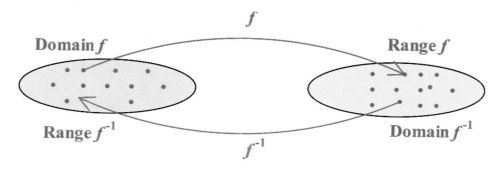

Existence of an Inverse Function

If the function has an inverse that is also a function, then there can only be one y for every x.

A **one-to-one** function is a function in which for every x there is exactly one y, and for every y, there is exactly one x. For $f(x)$ to have an **inverse function,** it must be **one-to-one.**

Some functions do not have inverse functions. For example, the function $f(x) = x^2$ is not a one-to-one function since there are two numbers that f takes to 1, $f(1) = 1$ and $f(-1) = 1$.

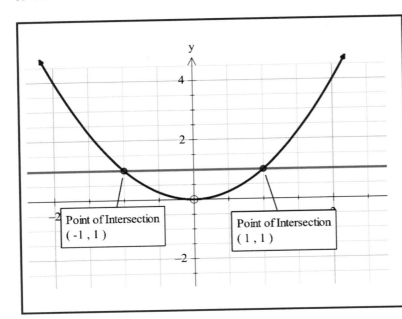

One way to check if a function is one-to-one is the **Horizontal Line Test.**

If a horizontal line intersects the graph of the function more than once, then the function is not one-to-one.

If no horizontal line intersects the graph of the function more than once, then the function is one-to-one.

For example, the function $f(x) = x^2$ is not one-to-one since line $y = 1$ intersects the graph of the function twice.

The variable y is said to be **inversely proportional** to x if $y = \frac{k}{x}$, where constant k $(k \neq 0)$ is called **constant of proportionality.** The variables, x and y, are in **inverse variation** $\left(y \propto \frac{1}{x}\right)$.

The variable y is said to be **directly proportional** to x if $y = kx$, where constant k $(k \neq 0)$ is called **constant of proportionality.** The variables, x and y, are in **direct variation** $(y \propto x)$.

Direct variation: $y = kx^n$ $(y \propto x^n)$

Inverse variation: $y = \frac{k}{x^n}$ $\left(y \propto \frac{1}{x^n}\right)$

Asymptotes

Horizontal Asymptote

A **Horizontal Asymptote** is a horizontal line that the graph of a function approaches, as x approaches negative or positive infinity. A horizontal asymptote may intersect the graph of the function.

For the following example (Graph 1), the function f has a horizontal asymptote at $y = -1$, and the function g has a horizontal asymptote at $y = 1$.

Vertical Asymptote

A **Vertical Asymptote** of a curve is a line of the form $x = a$ such that as x approaches some constant value a then the curve goes towards infinity (positive or negative). <u>A vertical asymptote doesn't intersect the graph of the function.</u>(Graph 2)

For the following example (Graph 2), the function f has a vertical asymptote at $x = 1$ and a horizontal asymptote at $y = 0$.

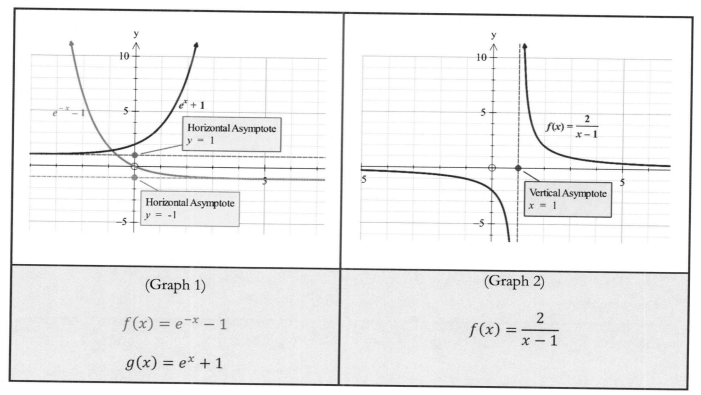

(Graph 1)	(Graph 2)
$f(x) = e^{-x} - 1$ $g(x) = e^x + 1$	$f(x) = \dfrac{2}{x - 1}$

Exponential Functions

An Exponential function is a function of the form $f(x) = a^x$ where a is a positive constant and $a \neq 1$.

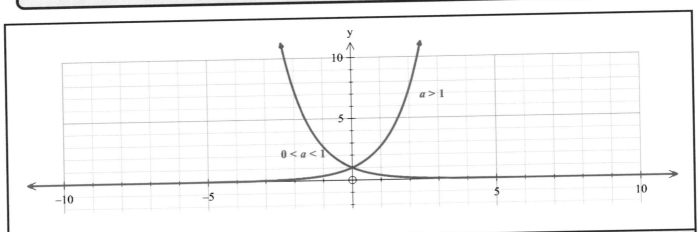

If $a > 1$, then the exponential function increases	If $0 < a < 1$, then the exponential function decreases

In either case, the x-axis is its horizontal asymptote

The domain of $f(x) = a^x$ consists of all real numbers, and its range consists of positive numbers only.

Note: We can use a GDC to solve exponential equations by drawing their graphs and finding the point(s) of intersection.

The general exponential functions are of the form:

$$f(x) = ka^x + c \, , a \in \mathbb{Q}^+, a \neq 1, k \neq 0$$
$$f(x) = ka^{-x} + c \, , a \in \mathbb{Q}^+, a \neq 1, k \neq 0$$

The equation of the horizontal asymptote is $y = c$

Logarithms

The inverse of the exponential function $f(x) = a^x$ is the logarithmic function with base a

$$f^{-1}(x) = \log_a x, \text{ where } a, x > 0 \text{ and } a \neq 1$$

The logarithm with base e is referred to as a **natural logarithm** $\log_e x \equiv \ln x$ where e is the Euler's number which is defined as $e = 2.71828..$

$$a^x = b \Leftrightarrow x = \log_a b, \text{ where } a, b > 0, a \neq 1$$

Trigonometry & Pythagoras' Theorem

▨ Circle Sectors and Segments

A **sector** of a circle is a region of a circle bounded by a central angle and its intercepted arc. The region of a circle bounded by an arc and a chord is called a **segment** of a circle.

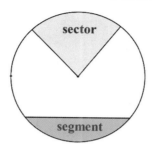

Arc length L	Sector Area
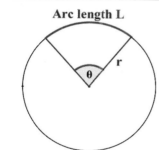	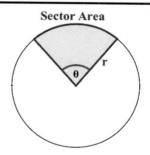
Arc Length: $L = \dfrac{\theta}{360} \times 2\pi r$ θ: the central angle in degrees, r: the radius	**Sector Area:** $A = \dfrac{\theta}{360} \times \pi r^2$ θ: the central angle in degrees, r: the radius
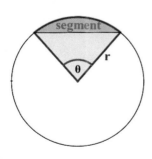	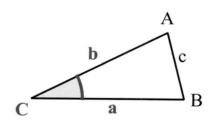
Area of segment = **Area of the sector - Area of the triangle=** $$\dfrac{\theta \pi r^2}{360} - \dfrac{1}{2}\, r^2 \sin\theta$$	**Two sides and the included angle** **Area of a triangle:** $A = \dfrac{1}{2}\, a\, b\, \sin C$

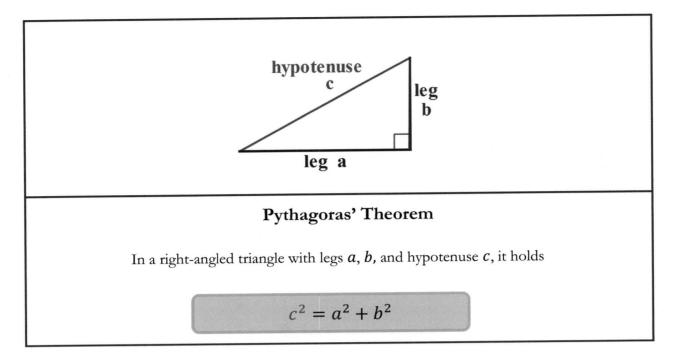

Pythagoras' Theorem

In a right-angled triangle with legs a, b, and hypotenuse c, it holds

$$c^2 = a^2 + b^2$$

Trigonometric ratios

For the following **right-angled triangle**, we have

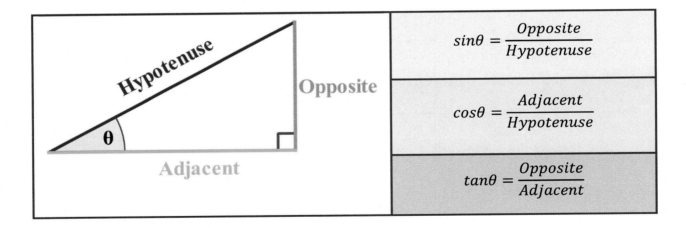

$$\sin\theta = \frac{Opposite}{Hypotenuse}$$

$$\cos\theta = \frac{Adjacent}{Hypotenuse}$$

$$\tan\theta = \frac{Opposite}{Adjacent}$$

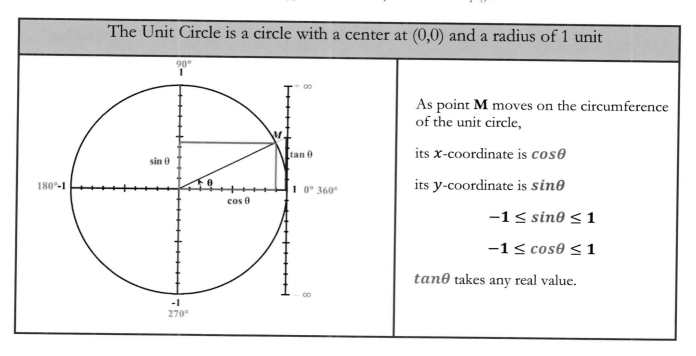

The Unit Circle is a circle with a center at (0,0) and a radius of 1 unit

As point **M** moves on the circumference of the unit circle,

its x-coordinate is $cos\theta$

its y-coordinate is $sin\theta$

$$-1 \leq sin\theta \leq 1$$

$$-1 \leq cos\theta \leq 1$$

$tan\theta$ takes any real value.

Sine rule

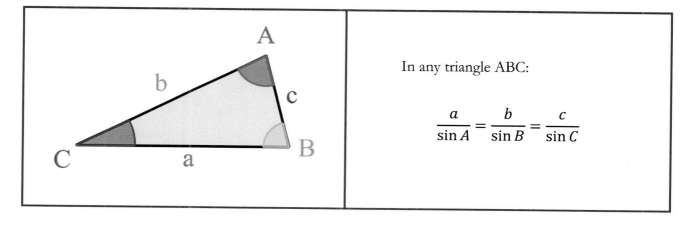

In any triangle ABC:

$$\frac{a}{\sin A} = \frac{b}{\sin B} = \frac{c}{\sin C}$$

Ambiguous case (optional): When you are given two adjacent sides of a triangle followed by an angle, the **sine rule** will provide you with two answers. The $sin^{-1}x$ function will only give us one of the two angles if we are using a calculator. To find the other one, we need to subtract the calculator's answer from 180°.

Cosine rule

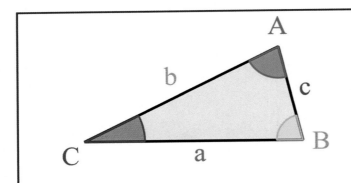

In any triangle, ABC

$$a^2 = b^2 + c^2 - 2\,b\,c\cos A$$
$$b^2 = c^2 + a^2 - 2\,c\,a\cos B$$
$$c^2 = a^2 + b^2 - 2\,a\,b\cos C$$

$$\cos A = \frac{b^2 + c^2 - a^2}{2bc} \qquad \cos B = \frac{c^2 + a^2 - b^2}{2ca} \qquad \cos C = \frac{a^2 + b^2 - c^2}{2ab}$$

The **bearing** to a point is the angle measured in a **clockwise** direction from the **north**.

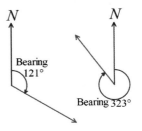

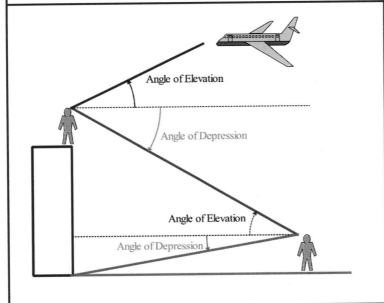

The **angle of elevation** denotes the angle from the horizontal upward to a point.

The **angle of depression** denotes the angle from the horizontal downward to a point.

Example

In a triangle ABC, $AB = 10\ cm$, $BC = 7\ cm$ and $\hat{B} = 40°$. Find the area of the triangle and the length of AC.

Solution

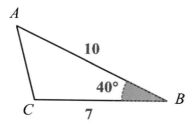

The area of the triangle is $A = \frac{1}{2} \times 7 \times 10 \times \sin 40° = 22.5\ cm^2$

By the **cosine rule**,

$$AC^2 = AB^2 + BC^2 - 2 \times AB \times BC \times \cos \hat{B}$$
$$AC^2 = 10^2 + 7^2 - 2 \times 10 \times 7 \times \cos 40°$$
$$AC^2 = 10^2 + 7^2 - 2 \times 10 \times 7 \times \cos 40°$$
$$AC^2 = 41.753 \Rightarrow AC = 6.46\ cm$$

▪ Functions that have the same general shape as a sine or cosine function are known as **sinusoidal functions**. The general forms of sinusoidal functions are:

$a\ sin(bx) + d$ $a\ cos(bx) + d$
Amplitude: $\|a\|$, Period: $\frac{360°}{b}$, principal axis: $y = d$

Example

Find the period, the amplitude, and the principal axis of the following trigonometric functions.

(i) $2sin(3x) + 3$

(ii) $-4sin\left(\frac{x}{2}\right) - 7$

Solution

(i) Period: $\frac{360°}{3} = 120°$, amplitude: 2, principal axis: $y = 3$

(ii) Period: $\frac{360°}{\frac{1}{2}} = 720°$, amplitude: 4, principal axis: $y = -7$

Trigonometric Functions

$f(x) = sinx$	The **Period** is the length that it takes for the curve to start repeating itself.
Domain: $x \in \mathbb{R}$, Range: $-1 \le y \le 1$ Period: $360°$	The **Amplitude** of a trigonometric function $(\sin x, \cos x)$ is the distance between the principal axis and one of the maximum or minimum points.

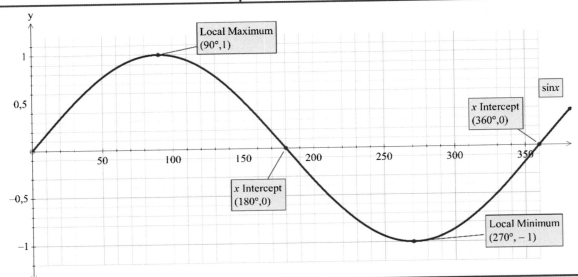

Amplitude: $\frac{y_{max}-y_{min}}{2} = \frac{1-(-1)}{2} = 1$	Principal axis: $y = \frac{y_{max}+y_{min}}{2} = \frac{1+(-1)}{2} = 0$ (the x-axis)

$f(x) = cosx$	The distance between the x-coordinates of the minimum and maximum of the graphs of $sinx$ or $cosx$ is half of a period.
Domain: $x \in \mathbb{R}$, Range: $-1 \le y \le 1$	Period: $360°$

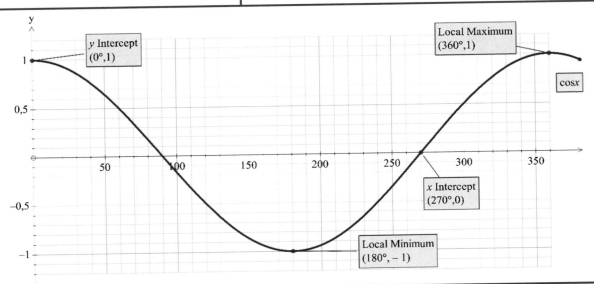

Amplitude: $\frac{y_{max}-y_{min}}{2} = \frac{1-(-1)}{2} = 1$	Principal axis: $y = \frac{y_{max}+y_{min}}{2} = \frac{1+(-1)}{2} = 0$ (the x-axis)

Differential Calculus

Derivative

The **tangent line** to the curve $y = f(x)$ at the point $A(x, f(x))$ is the line through A with slope (gradient) $f'(x)$ which is called the **derivative** of $f(x)$ at x.

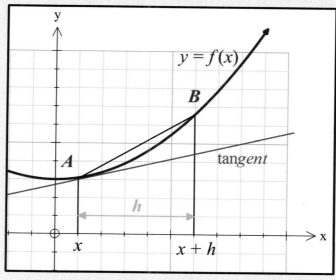

As point B approaches point A, h approaches 0, and the gradient m_{AB} of the line segment AB, $m_{AB} = \frac{f(x+h)-f(x)}{(x+h)-x}$, approaches the gradient of the tangent at point A. The line passes through the points A and B is called a **secant**.

The **rate of change** of the function f at $A(x, f(x))$ is given by the **gradient** of the **tangent** to the curve at A.

   The **gradient of the tangent** is the **limit of the secant**.

Apart from the **Newtonian notation (prime)** there is also the **Leibniz notation** for the derivative of the function $y = f(x)$

$$f'(x) = y' = \frac{dy}{dx} = \frac{df}{dx} = \frac{d}{dx}f(x)$$

$\frac{dy}{dx}$ measures the **rate of change** of y in respect of x.

Differentiation Rules

Function	Derivative
$f(x) = c$	$f'(x) = 0$
$f(x) = x$	$f'(x) = 1$
$f(x) = ax$	$f'(x) = a$
$f(x) = x^n, \quad n \in \mathbb{Z}$	$f'(x) = nx^{n-1}, \quad n \in \mathbb{Z}$
$f(x) = ax^n, \quad a \in \mathbb{R}, n \in \mathbb{Z}$	$f'(x) = nax^{n-1}, \quad a \in \mathbb{R}, n \in \mathbb{Z}$
$f(x) + g(x) = ax^n \pm bx^m$	$f'(x) + g'(x) = nax^{n-1} \pm mbx^{m-1}$

Examples

Find the derivative of the following functions

1) $f(x) = x^3 + 2x^2 - 3x + 5$

2) $g(x) = 3x^2 - \dfrac{5}{2x^6}$

Solution

Using the differentiation rules, we can differentiate sums of powers of x

1) $f'(x) = (x^3 + 2x^2 - 3x + 5)' = (x^3)' + (2x^2)' - (3x)' + (5)' = 3x^2 + 4x - 3 + 0 =$
$$= 3x^2 + 4x - 3$$

2) $g'(x) = \left(3x^2 - \dfrac{5}{2x^6}\right)' = \left(3x^2 - \dfrac{5}{2}x^{-6}\right)' = (3x^2)' - \left(\dfrac{5}{2}x^{-6}\right)' = 6x - (-6)\dfrac{5}{2}x^{-7} = 6x + \dfrac{15}{x^7}$

Example

Point A lies on the curve $f(x) = x^2 - 4x$. The gradient of the curve at A is equal to 6. Find the coordinates of point A.

Solution

By differentiating, the gradient function is $f'(x) = 2x - 4$. By setting $f'(x) = 6$, we can find the x-coordinate of A.

$$2x - 4 = 6$$

$$2x = 10$$

$$x = 5$$

Finally, to find the y-coordinate of A, we plug into the original function its x-coordinate.

$$f(5) = 5^2 - 4 \times 5 = 5$$

So, point A has coordinates $(5,5)$.

Example

Find the derivative of $f(x) = \frac{x^2 - 2x^5}{x^7}$

Solution

$$f'(x) = \left(\frac{x^2 - 2x^5}{x^7}\right)' = \left(\frac{x^2}{x^7} - \frac{2x^5}{x^7}\right)' = (x^{-5} - 2x^{-2})' =$$

$$= (x^{-5})' - (2x^{-2})' = -5x^{-6} - 2(-2)x^{-3} = -\frac{5}{x^6} + \frac{4}{x^3}$$

Equations of Tangent and Normal

> The equation of the **tangent** to a curve $y = f(x)$ at (x_1, y_1) is given by
> $$y - y_1 = f'(x_1)(x - x_1)$$
> Since the slope (m_T) of the tangent is $m_T = f'(x_1)$.

> The equation of the **normal** (the perpendicular to the tangent) to a curve $y = f(x)$ at (x_1, y_1) is given by
> $$y - y_1 = -\frac{1}{f'(x_1)}(x - x_1)$$
> Since the slope (m_N) of the normal is $m_N = -\frac{1}{m_T} = -\frac{1}{f'(x_1)}$

Note: When the first derivative (the slope of the tangent) at a certain point (x_1, y_1) is equal to zero $(f'(x_1) = 0)$ then the equation of the tangent at this point is given by the equation: $y = y_1$ (a horizontal line) and the equation of the normal at this point is a vertical line of the form $x = x_1$.

Example

Find the equations of the tangent and the normal to the curve $y = x^3$ at the point $(2,8)$.

Solution

$$f(x) = x^3$$
$$f'(x) = 3x^2$$
$$f'(2) = 3 \times 2^2 = 12$$

So, the slope of the tangent is $m_T = 12$, the slope of the normal is $m_N = -\frac{1}{12}$

moreover, the corresponding equations are given by the following formulas:

Equation of **tangent** at $(2,8)$: $y - 8 = 12(x - 2)$

Equation of **normal** at $(2,8)$: $y - 8 = -\frac{1}{12}(x - 2)$

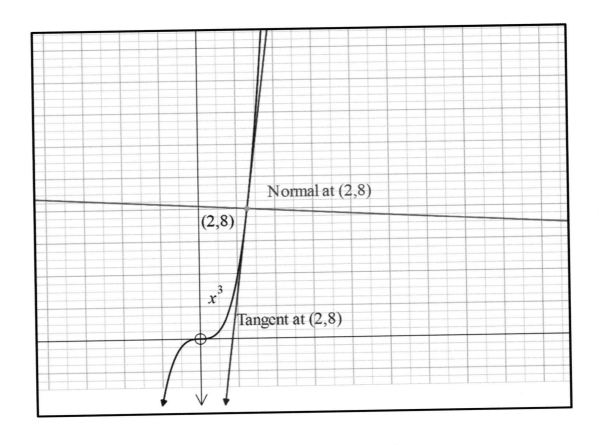

Stationary points

> A **stationary point** is a point where $f'(x) = 0$. It could be a local **minimum**, local **maximum**, or a **stationary point of inflection**.

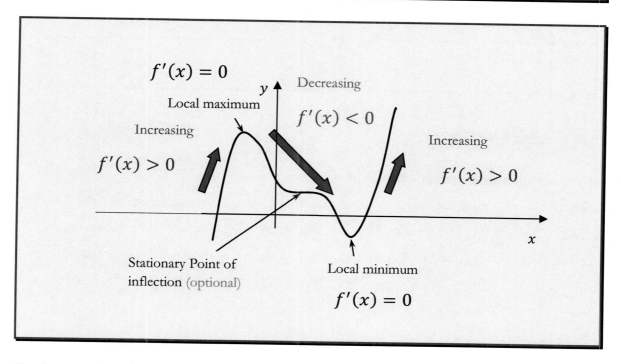

Test for increasing/decreasing

- If $f'(x) \geq 0$ on an interval, then f is **increasing** on this interval.

- If $f'(x) \leq 0$ on an interval, then f is **decreasing** on this interval.

The First Derivative Test for turning points (maximum/minimum)

Suppose x_0 is a **stationary point** ($f'(x_0) = 0$) of a continuous function f.

- If f' changes from positive to negative at x_0, then f has a local **maximum** at x_0.

		x_0	
f'	+	0	-
f	↗	max	↘

■ If f' changes from negative to positive at x_0, then f has a local **minimum** at x_0.

			x_0		
f'		-	0		+
f		↘	min		↗

Optimization Problems

Many application problems in calculus involve functions for which you want to find maximum or minimum values. First, we have to write down the "constraint" equation and the "optimization" equation. Then, we have to express the optimization equation as a function of only one variable, and, if it is needed, reduce it to be easily differentiable. Finally, we have to differentiate the function and set it equals to zero to find the stationary points. If we have a closed interval, apart from the stationary points, we should also examine the endpoints for maximum or minimum.

Example

A closed cylindrical tin is to be made from a sheet of metal measuring $600\pi \ cm^2$. Find the dimensions of the tin if the volume is to be maximum.

Solution

The total surface area of a closed cylinder is

$$A = 2\pi r^2 + 2\pi rh = 600\pi$$

$$h = \frac{600\pi - 2\pi r^2}{2\pi r} = \frac{600 - 2r^2}{2r} = \frac{300 - r^2}{r}$$

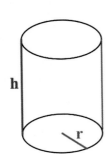

and the volume is $V = \pi r^2 h = \pi r^2 \frac{300-r^2}{r} = 300\pi r - \pi r^3$

$$V'(r) = (300\pi r - \pi r^3)' = 300\pi - 3\pi r^2 = 0$$

$$r = \sqrt{\frac{300}{3}} = \sqrt{100} = 10$$

Since $V'(r) > 0$ when $r < 10$ and $V'(r) < 0$ when $r > 10$, we can conclude that

the volume is maximized when $r = 10 \ cm$ and $h = \frac{600 - 2 \times 10^2}{2 \times 10} = \frac{600 - 200}{20} = 20$ cm

Integration

Indefinite Integral

If $F(x)$ is a function where $F'(x) = f(x)$ then the antiderivative of $f(x)$ is $F(x)$ and the indefinite integral is defined as

$$\int f(x)dx = F(x) + c$$

Properties of Indefinite Integrals

- $\int f'(x)dx = f(x) + c$

- $\left(\int f(x)dx\right)' = f(x)$

- $\int kf(x)dx = k\int f(x)dx$, where k is any constant.

- $\int(f(x) \pm g(x))dx = \int f(x)dx \pm \int g(x)dx$

Basic Indefinite Integrals

$$\int kdx = kx + c$$

$$\int x^n dx = \frac{x^{n+1}}{n+1} + c$$

$$(n \neq -1)$$

Example

If $\frac{dy}{dx} = 12x^3 + 6x^2 - 5$ and $y = 7$ when $x = -1$, find y in terms of x.

Solution

$$y = \int(12x^3 + 6x^2 - 5)dx = \frac{12x^4}{4} + \frac{6x^3}{3} - 5x + c = 3x^4 + 2x^3 - 5x + c$$

$$y = 3x^4 + 2x^3 - 5x + c$$

$$7 = 3(-1)^4 + 2(-1)^3 - 5(-1) + c \Rightarrow c = 1$$

So, $y = 3x^4 + 2x^3 - 5x + 1$

Definite Integral

A definite integral is of the form

$$\int_a^b f(x)\,dx$$

, where x is called the variable of integration and a, b are called the **Lower** and **Upper limit** respectively.

Example

Find the value of the following definite integral by using a GDC

$\int_0^2 (12x^3 + 6x^2 - 5)\,dx$

Solution

$$\int_0^2 (12x^3 + 6x^2 - 5)\,dx = 54$$

TI 84 + (Definite Integral)		Casio fx9860 series, fx-CG20, fx-CG50 (Definite Integral)
fnInt	Press **MATH** Press ▼ several times to select **9: fnInt(** Press **ENTER** and then fill out all the required information.	Press **MENU** **Run-Matrix** **EXE** then select **MATH (F4)** **▶(F6)** **F1** and then fill out all the required information.

Applications of Integration

Areas above the x-axis

For a function $f(x) \geq 0$ on an interval $[a, b]$, the area between the x -axis and the curve $y = f(x)$ between $x = a$ and $x = b$, is given by

$$A = \int_a^b f(x)\,dx$$

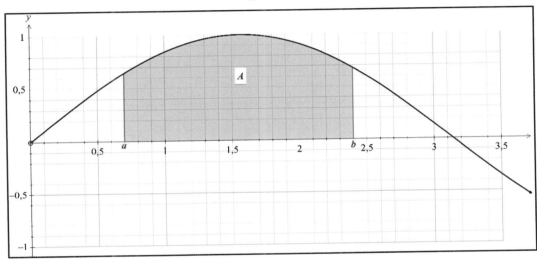

The Trapezoidal rule

Trapezoidal rule is used to find an approximation of a definite integral. The fundamental concept in the trapezoidal rule is to split the region under the graph of the given function into several trapezoids, as shown in the figure below, and then we can easily evaluate the sum of their areas.

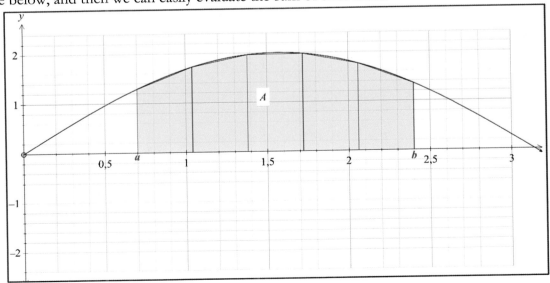

By adding the area of the n trapezoids, we obtain the following approximation:

$\int_a^b y \, dx \approx \frac{1}{2} h \big((y_0 + y_n) + 2(y_1 + y_2 + \cdots + y_{n-1}) \big)$, in which the interval $[a, b]$ is

divided into n subintervals of equal length h, where $h = \frac{b-a}{n}$.

Example

Using the trapezoidal rule, with all values of the following table, find an approximation for the area under the curve over the interval $[1,2]$.

x	1	1.2	1.4	1.6	1.8	2
y	1	1.44	1.96	2.56	3.24	4

$$\int_1^2 y \, dx \approx \frac{1}{2} \times \frac{1}{5} \times \big((1 + 4) + 2(1.44 + 1.96 + 2.56 + 3.24) \big) = 2.34$$

Statistics

Population: The entire group you want to know something about.

Sample: The group you use to infer something about the population.

A **random sample** is a set of n objects in a population of N objects where all possible samples are equally likely to happen.

Simple random sampling: Every member of the population has an equal chance of being selected into the sample.

Convenience sampling entails using the most conveniently available members of being selected into the sample.

Systematic sampling is a sampling method used to pick a sample of size n from a target population of size N utilizing a predetermined system. In systematic sampling, every kth member of the population is selected to be included in the study, where $k = \frac{N}{n}$, rounded down to the nearest whole number.

Stratified sampling is a technique in which the entire population is divided into different strata (subgroups), and then we randomly select the sample's members proportionally from the various strata.

Quota sampling is somewhat similar to stratified sampling. Instead of choosing randomly from strata within the population, quota sampling involves dividing a population into various categories and then setting quotas as to how many elements to select (non-random) from each category based on certain criteria.

Sampling Error is a statistical error that arises when a researcher fails to pick a sample representing the entire data population, and the results contained in the sample do not reflect the results that should be obtained from the entire population.

Continuous data can be assigned an infinite number of values between whole numbers (e.g., a person's height or weight).

Discrete data is data that can be counted (e.g., the number of students).

A **Bar chart** is a graph that uses vertical or horizontal bars to represent the frequencies of the categories in a data set. (Categorical variables)

A **Histogram** is a graphical display of a frequency or a relative frequency distribution that uses classes and vertical bars (without any spaces between them) of various heights to represent the frequencies (Quantitative variables). Histograms are used for displaying variable distributions, while bar charts are used for comparing variables. Histograms plot quantitative data with information ranges grouped in intervals, while bar charts plot categorical data.

A **frequency polygon** is a graph that displays the data using lines to connect points plotted for the frequencies. The frequencies represent the heights of the vertical bars in the histograms.

A **stem and leaf plot** is a table where all the data must be first sorted in ascending order, and then each data value is split into a "stem" (the first digit or digits) and a "leaf" (usually the last digit).

Measures of Central Tendency

The **mean** of a set of values is the number obtained by adding the values and dividing the total by the number of values. In the formula below x_i represents an observation of the data, f_i the corresponding frequency and n the total number of observations.

$$\bar{x} = \frac{1}{n} \sum_{i=1}^{k} f_i x_i = \frac{f_1 x_1 + f_2 x_2 + \cdots + f_k x_k}{n}$$

$$n = \sum_{i=1}^{k} f_i = f_1 + f_2 + \cdots + f_k$$

The **population mean** is usually denoted by μ and the **sample mean** is denoted by $\bar{x}$ (read as 'x-bar').

Notes:

- The calculation of the mean can be performed by using the formula or technology.
- When the data are grouped into classes, we should use the midpoint or mid-interval value to represent all values within that interval in order to estimate the mean of grouped data.

Example

Find the mean of a sample of 6 test grades $(80, 65, 91, 75, 82, 76)$.

Answer

$$\bar{x} = \frac{\sum_{i=1}^{6} x_i}{6} = \frac{80 + 65 + 91 + 75 + 82 + 76}{6} \cong 78.17$$

Example (Grouped Data)

Estimate the mean and write down the modal class of the following heights

Heights (cm)	158-160	161-163	164-166	167-169	170-172	173-175	176-178	179-181
Mid-height x_i (cm)	159	162	165	168	171	174	177	180
Frequency f_i	2	4	4	5	7	6	3	1

From the table above, we have that

$$\sum_{i=1}^{8} f_i = 2 + 4 + 4 + 5 + 7 + 6 + 3 + 1 = 32$$

and

$$\sum_{i=1}^{8} f_i x_i = 159 \times 2 + 162 \times 4 + 165 \times 4 + 168 \times 5 + 171 \times 7 + 174 \times 6 + 177 \times 3 + 180 \times 1$$

$$= 5418$$

$$\bar{x} = \frac{\sum f_i x_i}{\sum f_i} = \frac{5418}{32} = 169.31$$

The **modal class** is $170 - 172$ since it has the largest frequency (7).

▣ The **median** of a data set is the middle value when the data values are arranged in ascending or descending order. If the data set has an even number of entries, the median is the mean of the two middle data entries.

Examples

1. Sample $(8, 3, 5, 12, 15, 20, 1), n = 7$

Answer

Step 1: arrange in ascending order $1, 3, 5, \mathbf{8}, 12, 15, 20$

Step 2: the median is 8

2. Sample $(8, 3, 5, 12, 15, 20, 1, 13), n = 8$

Answer

Step 1: arrange in ascending order $1, 3, 5, \mathbf{8}, \mathbf{12}, 13, 15, 20$

Step 2: the median (m) is given by $m = \frac{8+12}{2} = 10$

▣ The **mode** of a data set is the value that occurs most frequently. When two values occur with the same greatest frequency, each one is a mode, and the data set is **bimodal**. When more than two values occur with the same greatest frequency, each is a mode, and the data set is said to be **multimodal**. When no value is repeated, we say there is no mode.

Example

Find the mode of a sample of 7 test grades $(80, 65, 91, 75, 82, 76, 82)$.

Answer

The **mode** is grade 82 because it has the greatest frequency $(= 2)$

- In a **negatively skewed distribution**, the mean is to the left of the median, and the mode is to the right of the median.

$$\text{Mean<Median<Mode}$$

- In a **symmetrical distribution**, the data values are evenly distributed on both sides of the mean. Also, when the distribution is unimodal, the mean, median, and mode are all equal and are located at the center of the distribution.

$$\text{Mean=Median}$$

- In a **positively skewed distribution**, the mean is to the right of the median, and the mode is to the left of the median.

$$\text{Mode<Median<Mean}$$

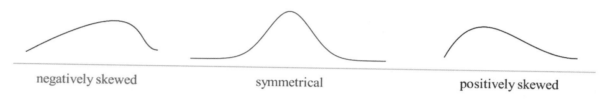

| negatively skewed | symmetrical | positively skewed |

Measures of Dispersion

Range=largest measurement-smallest measurement

Example

Sample $(12, 15, 18, 20, 17)$

$Range = max - min = 20 - 12 = 8$

Variance	Standard Deviation
$$s_n^2 = \frac{1}{n}\sum_{i=1}^{n}(x_i - \bar{x})^2$$	$$s_n = \sqrt{\frac{1}{n}\sum_{i=1}^{n}(x_i - \bar{x})^2}$$

Notes:

- Standard deviation is the square root of variance.
- Calculation of standard deviation and variance using only technology.

Percentiles

Let $0 < p < 100$. The pth percentile is a number x such that $p\%$ of all measurements fall below the pth percentile and $(100 - p)\%$ fall above it.

Lower Quartile (25th percentile) Q_1	Median (50th percentile) Q_2	Upper Quartile (75th percentile) Q_3

The **Interquartile range (IQR)** of a data set is the difference between the third and first quartile.

$$IQR = Q_3 - Q_1$$

Box-and-whisker plot

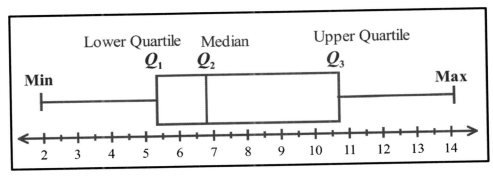

- The ends of the box are the **upper** (Q_3) and **lower** (Q_1) quartiles, so the box spans the **interquartile range** (IQR)

- The **median** (Q_2) is marked by a vertical line inside the box.

- The **whiskers** are the two lines outside the box that extend to the maximum and minimum observations.

- An **outlier** is a point that falls more than **1.5** times the interquartile range ($1.5 \times IQR$) above the third quartile or below the first quartile.

- The **IQR** is often the preferred measure of spread since it is not affected by outliers.

Important: A GDC may be used to produce **histograms** and **box-and-whisker plots**.

- If we **add** or **subtract** a positive constant value c to/from all the numbers in the dataset, the **mean** (and the **median**) will also **increase** or **decrease**, respectively, by c and the **standard deviation** (and the **variance**) will remain the **same**.

- If we **multiply** or **divide** all the numbers in the dataset by a positive constant value c, the **mean**, the **median** and the **standard deviation** will be **multiplied** or **divided**, respectively, by c. It follows that the **variance** will be **multiplied** or **divided** by c^2.

Example

A data set has a mean of **24** and a standard deviation of **4**.

(a) Each value in the data set has **7** added to it. Find the new mean, the new standard deviation, and the new variance.

(b) Each value in the data set is multiplied by **5**. Find the new mean, the new standard deviation, and the new variance.

Answer

(a) The new mean is $24 + 7 = 31$, and both the standard deviation and variance remain the same.

(b) The new mean is $24 \times 5 = 120$, the new standard deviation is $4 \times 5 = 20$, and the new variance is $20^2 = 400$.

Example of **Histogram** and **Cumulative frequency diagram**

The following cumulative frequency table displays the marks obtained in a test by a group of 80 students. The cumulative frequency is calculated by accumulating the frequencies as we move down the table.

Grades	Frequency	Cumulative frequency
[0-20]	5	5
(20-40]	10	15
(40-60]	25	40
(60-80]	25	65
(80-100]	15	80

The corresponding Histogram is presented below:

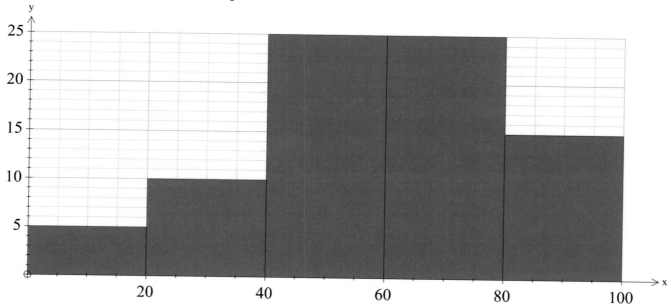

The corresponding cumulative frequency diagram is presented below:

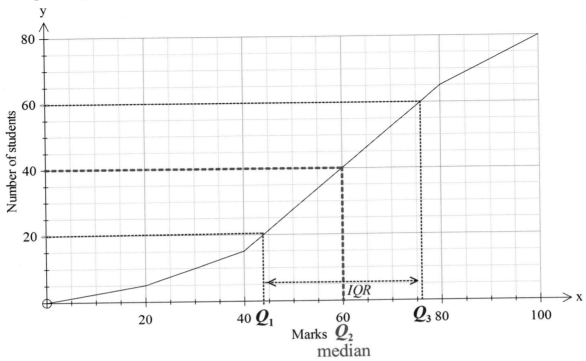

The **median** is estimated using the 50th percentile. As 50% of 80 is 40, we draw a horizontal line parallel to x-axis (Marks), passing through 40 until this line cuts the curve. Then we draw a vertical line parallel to y-axis (Number of students) until it reaches x-axis; in this case, the value is 60 marks. Similarly, the lower and upper quartiles Q_1, Q_3 can be found at a cumulative frequency of 20 (25th percentile) and 60 (75th percentile) number of students, respectively. Following the above steps, we get $Q_1 = 44$ and $Q_3 = 76$. Thus, the interquartile range is $IQR = Q_3 - Q_1$.

Probability

- A **sample space** U (Universal Set) is the set of all possible outcomes of an experiment.

- An **event** is one or more outcomes of an experiment. Mathematically, an event is a subset of a sample space. For example, scoring a four on the throw of a die.

- An **event** is **simple** if it consists of just a single outcome and is **compound** otherwise.

- A sample space is discrete if it consists of a finite or countable infinite set of outcomes. A sample space is continuous if it contains an interval (either finite or infinite) of real numbers.

- The set containing no elements is called an **empty set** (or null set) and denote it by $\emptyset$.

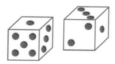

Example

If we toss a coin three times and record the result, the sample space is

$$U = \{HHH, HHT, HTH, HTT, THH, THT, TTH, TTT\}$$

where (for example) THH means 'Tails on the first toss, then heads, then heads again.'

The **theoretical probability** of an occurring event A is given by

$$P(A) = \frac{n(A)}{n(U)} = \frac{Number\ of\ outcomes\ in\ which\ A\ occurs}{Total\ number\ of\ outcomes\ in\ the\ sample\ space\ U}$$

Axioms of probability

1. $0 \le P(A) \le 1$

2. $P(\emptyset) = 0$ and $P(U) = 1$

3. If A and B are both subsets of U and they are **mutually exclusive** ($A \cap B = \emptyset$), then

$$P(A \cup B) = P(A) + P(B)$$

We also have the following propositions:

- $P(A') = 1 - P(A)$, where A' is the complement event of A

- If $A \subseteq B$ then $P(A) \leq P(B)$

- $P(A \cup B) = P(A) + P(B) - P(A \cap B)$

Example

A six-sided die is rolled twice. What is the probability that the sum of the numbers is at least 10?

Answer

The number of elements in the sample space is $6^2 = 36$. To obtain a sum of 10 or more, the possibilities for the numbers are $(4,6), (5,5), (6,4), (5,6), (6,5),$ or $(6,6)$. So the probability of the event "that the sum of the numbers is at least 10" is $\frac{6}{36} = \frac{1}{6}$.

Example

A box contains 4 red and 5 blue disks. A disk is randomly selected and has its color noted. The disk is not replaced, and a second disk is then selected.

(a) Find the probability that the disks will be of a different color.
(b) Find the probability that they will be both red.

Answer

The tree diagram for this information is:

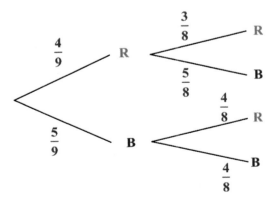

(a) $P(different\ color) = P(RB) + P(BR) = \frac{4}{9} \times \frac{5}{8} + \frac{5}{9} \times \frac{4}{8} = \frac{40}{72}$

(b) $P(both\ red) = P(RR) = \frac{4}{9} \times \frac{3}{8} = \frac{12}{72}$

Venn Diagrams

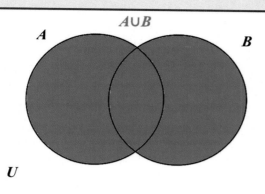

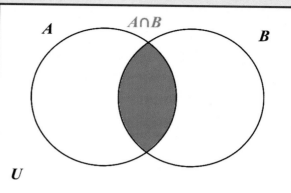

The **Union** of A and B is the set whose elements are all the elements of A and B.	The **Intersection** of A and B is the set consisting of the elements that belong to both A and B.
$$A \cup B = \{x : x \in A \ or \ x \in B\}$$ $$P(A \cup B) = P(A) + P(B) - P(A \cap B)$$	$$A \cap B = \{x : x \in A \ and \ x \in B\}$$

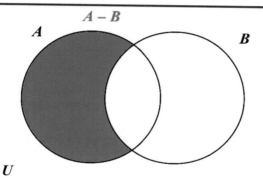

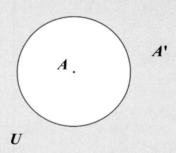

The **difference** of A and B is the set consisting of those elements of A that are not in B.	The **complement** of A is the set consisting of those elements of U that are not in A.
$$A - B = \{x : x \in A \ and \ x \notin B\}$$ $$P(A - B) = P(A) - P(A \cap B)$$	$$A' = A^c = U - A = \{x \in U : \ x \notin A\}$$ $$P(A') = 1 - P(A)$$

	If $A \cap B = \emptyset$ the events A and B are said to be **disjoint** or **mutually exclusive**. Disjoint sets are sets that do not have elements in common.
A B U	$$P(A \cup B) = P(A) + P(B)$$

Conditional Probability

The probability of an event given that another event has already occurred is a conditional probability.

If A and B are two events, then the **conditional probability** of event A **given (/)** an event B can be found by the formula

$$P(A/B) = \frac{P(A \cap B)}{P(B)}, P(B) \neq 0$$

It should also be noted that usually $P(A/B) \neq P(B/A)$

Example

A lot of **1000** semiconductor chips contains **40** that are defective. Two are selected randomly, without replacement, from the lot.

(a) What is the probability that the first one selected is defective?

(b) What is the probability that the second one selected is defective, **given** that the first one was defective?

(c) What is the probability that both are defective?

Solution

Let A: The first one selected is defective

and B: The second one selected is defective

Then

(a) $P(A) = \frac{40}{1000} = 0.04$

(b) $P(B/A) = \frac{P(B \cap A)}{P(A)} = \frac{\frac{40}{1000} \times \frac{39}{999}}{\frac{40}{1000}} = \frac{39}{999}$

(c) $P(B \cap A) = \frac{40}{1000} \times \frac{39}{999} = \frac{156}{99900}$

Independence

Two events A and B, are said to be **independent** if

$$P(A \cap B) = P(A) \times P(B)$$

alternatively, $P(A/B) = P(A)$ and $P(B/A) = P(B)$

Properties of independence

▪ If A and B are independent then A and B' are independent.
▪ If A and B are independent, so are A' and B'.

Note: A common mistake is to confuse whether two events are **independent** or **mutually exclusive**. A and B are mutually exclusive events or disjoint if $P(A \cap B) = 0$, that is, the occurrence of one precludes that of the other.

Examples

1. If $P(A/B) = 0.5, P(B) = 0.7$ and $P(A) = 0.4$, are the events A and B independent?

Answer

The events are not independent because $P(A/B) \neq P(A)$.

2. Given that $P(A) = 0.7, P(B) = 0.5$ and that A and B are **independent** events. Find the probability of the following events:

(a) $A \cap B$ **(b)** $A \cup B$ **(c)** A/B' **(d)** $A' \cap B$

Solution

The events are independent, therefore

(a) $P(A \cap B) = P(A) \times P(B) = 0.7 \times 0.5 = 0.35$

(b) $P(A \cup B) = P(A) + P(B) - P(A \cap B) = 0.7 + 0.5 - 0.35 = 0.85$

(c) $P(A/B') = \frac{P(A \cap B')}{P(B')} = \frac{P(A) - P(A \cap B)}{1 - P(B)} = \frac{0.7 - 0.35}{1 - 0.5} = \frac{0.35}{0.5} = 0.7$

Alternatively, we know that "If A and B are independent then A and B' are also independent", so

$$P(A/B') = P(A) = 0.7$$

(d) $P(A' \cap B) = P(A') \times P(B) = 0.3 \times 0.5 = 0.15$

Discrete Probability Distributions

A **random variable** is called **discrete** if it has either a finite or a countable number of possible values.

A **discrete probability distribution** describes the probability of occurrence of each value of a discrete random variable.

If X is a **discrete random variable** with $P(X = x_i)$, $i = 1,2, \dots n$ then

1. $0 \leq P(X = x_i) \leq 1$ for all values of x_i
2. $\sum_{i=1}^{n} P(X = x_i) = P(X = x_1) + P(X = x_2) + \cdots + P(X = x_n) = 1$
3. The **expectation** of the random variable X is

$$E(X) = \mu = \sum_{i=1}^{n} x_i P(X = x_i)$$

4. The **variance** is defined by the following formula

$$Var(X) = E(X^2) - [E(X)]^2$$

Note: If we have a **fair game,** then $E(X) = 0$ where X represents the gain of one of the players.
- For a discrete random variable X, the **mode** is any value of x with the highest probability and it may not be unique.

Example

The following table shows the probability distribution of a discrete random variable X.

x	0	2	4
$P(X = x)$	0.2	$3m$	m

a) Find the value of m
b) Find the expected value of X.

Solution

a) $\sum_{i=1}^{3} P(X = x_i) = 0.2 + 3m + m = 4m + 0.2 \xrightarrow{\text{the sum of probabilities equals 1}} 4m + 0.2 = 1$

$\Rightarrow 4m = 0.8 \Rightarrow m = 0.2$

b) $E(X) = \sum_{i=1}^{n} x_i P(X = x_i) = (0 \times 0.2) + (2 \times 0.6) + (4 \times 0.2) = 2$

Binomial Distribution

The **Binomial distribution** is a discrete probability distribution. It describes the outcome of n **independent** trials. Each trial is assumed to have only two outcomes, either **success** or **failure**. The probability of a success, denoted by p, remains **constant** from trial to trial. The probability of having exactly r **successes** in n independent trials is denoted as

$$P(X = r), where\ r = 0,1, \dots n$$

If a discrete random variable X follows a **Binomial distribution** with parameters n and p, $X \sim B(n, p)$, then the **mean** and the **variance** are given by the following formulas:

Expected value (mean)	Variance
$E(X) = np$	$Var(X) = np(1 - p)$

Example

A **fair** coin is tossed **seven** times. Calculate the probability of obtaining:

(i) Exactly four heads.

(ii) At least two heads.

Solution

(i) Let X denote the number of heads, $X \sim B(7,0.5)$

$$P(X = 4) = \binom{7}{4}(0.5)^4(0.5)^{7-4} = 0.273\ (3\ s.f.)$$

(ii) Let X denote the number of heads, $X \sim B(7,0.5)$

$$P(X \geq 2) = 1 - P(X \leq 1) = 1 - \big(P(X = 0) + P(X = 1)\big) = 0.938(3\ s.f.)$$

, where $P(X = 0) = \binom{7}{0}(0.5)^0(0.5)^{7-0} = 0.00781(3\ s.f.)$

$$P(X = 1) = \binom{7}{1}(0.5)^1(0.5)^{7-1} = 0.0547(3\ s.f.)$$

Note: The formulas used in the example above are not required. Binomial probabilities should be found using a GDC.

TI 84 + (Example)		Casio fx9860 series, fx-CG20, fx-CG50 (Example)	
$X \sim B(7, 0.5)$ $P(X = 4)$	**binompdf**($numtrials, p, x$) Computes a probability at x for the discrete binomial distribution with the specified $numtrials$ and probability p of success on each trial. 2nd [DISTR] **DISTR** **A:binompdf(7,0.5,4)**	$X \sim B(7, 0.5)$ $P(X = 4)$	On the initial STAT mode screen→F5(DIST)→F5(BINM)→F1(Bpd) we set Data: **Variable**, x: **4**, Numtrial: **7**, p: **0.5**→Execute P(X = 4) = 0.273 (3 s.f.)
$X \sim B(7, 0.5)$ $P(X \leq 1)$	**binomcdf**($numtrials, p, x$) Computes a cumulative probability at x for the discrete binomial distribution with the specified $numtrials$ and probability p of success on each trial. 2nd [DISTR] **DISTR** **B:binomcdf(7,0.5,1)**	$X \sim B(7, 0.5)$ $P(X \leq 1)$	On the initial STAT mode screen→F5(DIST)→F5(BINM)→F2(Bcd) we set Data: **Variable**, x: **1**, Numtrial: **7**, p: **0.5**→Execute P(X ≤ 1) = 0.063 and then P(X ≥ 2) = 1 − P(X ≤ 1) = 0.938(3 s.f.)

Note: Casio models fx-CG20, fx-CG50 calculate directly the probability $P(X \geq 2)$.

Normal Distribution

The **normal distribution** is a theoretical ideal distribution. Real-life empirical distributions never match this model perfectly. However, many things in life do approximate the normal distribution and are said to be "normally distributed".

The normal distribution has the following properties:

▦ Its shape is symmetric about the **mean** (μ), which is also the **median** and the **mode** of the distribution.

▦ It is a bell-shaped curve, with tails going down and out to the left and right, and the x −axis is a horizontal asymptote.

▦ Its standard deviation (σ), measures the distance on the distribution from the mean to the inflection point.

▦ The total area under the curve is equal to **1**.

▦ Approximately **68** percent of its values lie within one standard deviation of the mean.

▦ Approximately **95** percent of its values lie within two standard deviations of the mean.

▦ Approximately **99.7** percent of them lie within three standard deviations of the mean.

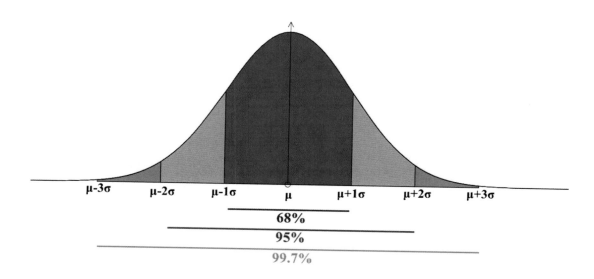

Example

The weights of a group of men are normally distributed with a mean of 80 kg and a standard deviation of 15 kg.

(i) A man is chosen at random. Find the probability that the man's weight is greater than 90 kg.

(ii) In this group, 20% of men weigh less than w kg. Find the value of w.

Solution

The required probability for **(i)** is represented in the following diagram.

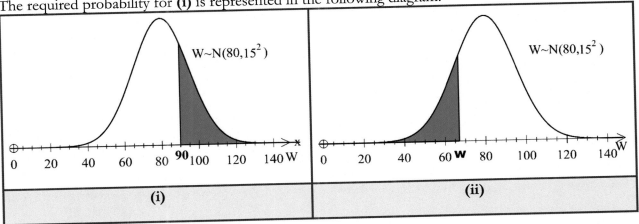

(i)	(ii)

In order to find this probability, we can use a GDC.

　▨　For **Casio fx9860 series, fx-CG20, fx-CG50**, we perform the following commands:

On the initial STAT mode screen→F5(DIST)→F1(NORM)→F2(Ncd)

we set Lower: **90**, Upper: **10^99**, σ: **15**, μ: **80**→ Execute

$$P(W > 90) = 0.252 \ (3 \ \text{s.f.})$$

　▨　For **TI 84 +**

2nd→VARS→2→Enter four parameters (lower limit: **90**, upper limit: **10^99**, mean: **80**, standard deviation: **15**)

(ii) To find the value of w, we have to use the inverse normal, which gives us an x-value if we input the area (probability region) to the left of the x-value.

To find this probability, we can use a GDC.

　▨　For **Casio fx9860 series, fx-CG20, fx-CG50**

On the initial STAT mode screen→F5(DIST)→F1(NORM)→F3(InvN)

we set Tail: Left, Area(probability less than): **0.20**, σ: **15**, μ: **80**→Execute

　▨　For **TI 84 +**

2nd→VARS→3→Enter three parameters (probability less than - area: **90**, mean: **80**, standard deviation: **15**)

$$P(W < w) = 0.20 \Rightarrow w = 67.4$$

Correlation - Regression

Correlation

Bivariate data is data that contains two variables. The relationship between these two variables is often represented by a scatter plot. The **bivariate analysis** explores the relationship between two variables, whether there is an association and the strength of this association.

Correlation describes the degree of relationship between two variables.

The most common measure of correlation in statistics is **Pearson's product-moment correlation coefficient**. It shows the **linear relationship** between two sets of data. It is represented by r and it has a range of

$$-1 \leq r \leq 1$$

The sign of r indicates the direction of the correlation.

$r > 0$ means a **positive** correlation

$r < 0$ means a **negative** correlation

The magnitude of r indicates the strength of the correlation.

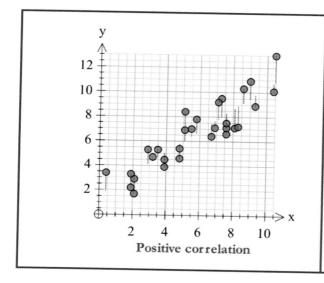

Positive correlation

A positive correlation is a relationship between two variables, where if one variable increases, the other one also increases. If the independent variable (x) increases then the dependent variable (y) also increases.

$$r > 0$$

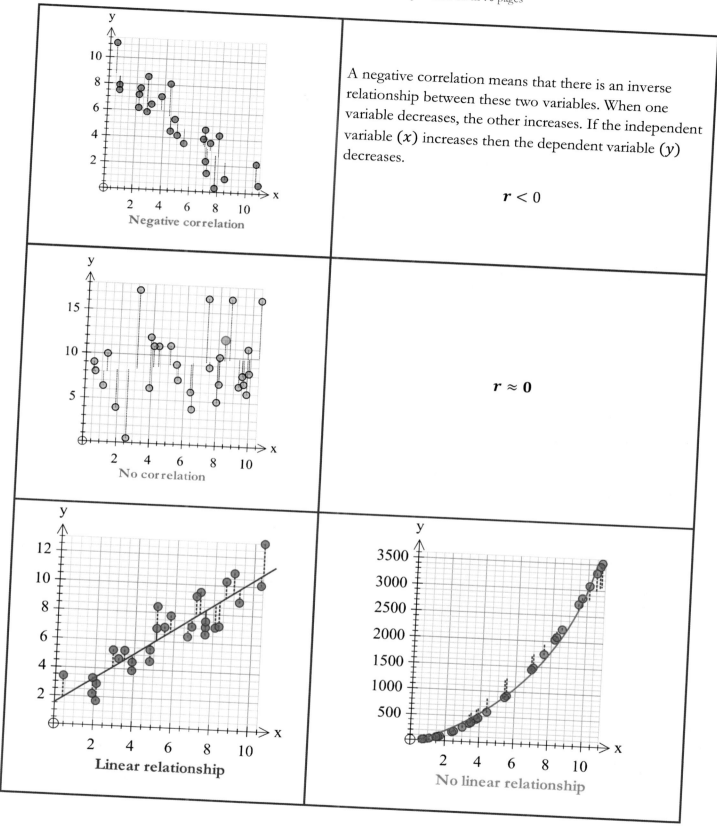

A negative correlation means that there is an inverse relationship between these two variables. When one variable decreases, the other increases. If the independent variable (x) increases then the dependent variable (y) decreases.

$$r < 0$$

$$r \approx 0$$

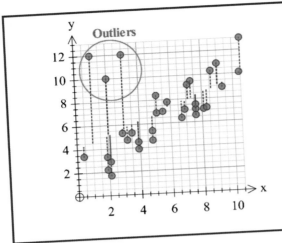

An outlier is an observation that lies an abnormal distance from other values in a random sample from a population.

Positive correlation		Negative correlation	
$r = 1$	perfect positive	$r = -1$	perfect negative
$0.95 \leq r < 1$	very strong positive	$-1 \leq r < -0.95$	very strong negative
$0.75 \leq r < 0.95$	strong positive	$-0.95 \leq r < -0.75$	strong negative
$0.50 \leq r < 0.75$	moderate positive	$-0.75 \leq r < -0.50$	moderate negative
$0.10 \leq r < 0.50$	weak positive	$-0.50 \leq r < -0.10$	weak negative
$0 \leq r < 0.10$	no correlation	$-0.10 < r \leq 0$	no correlation

Line of best fit

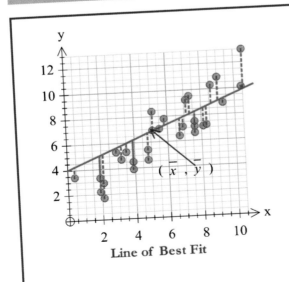

Line of Best Fit

A **line of best fit** is a straight line of the form

$$y = ax + b$$

that is the best approximation of the given set of data.

A **line of best fit** can be roughly determined by drawing a straight line on a scatter plot so that the number of points above the line and below the line is about equal, the line passes through as many points as possible and passes through the mean point $(\bar{x}, \bar{y})$.

Note: The more the correlation coefficient (r) is closer to 1 or -1, the more accurate the predictions produced by the line of best fit are.

Interpolation

We could use the **line of best fit** to predict the value of the **dependent variable (y)** for an **independent variable (x)** that is **within the range** of our data. In this case, we are performing **interpolation**.

Extrapolation

We could use the **line of best fit** to predict the value of the dependent variable for an independent variable that is **outside the range** of our data. In this case, we are performing **extrapolation**.

Of the two methods, **interpolation** is preferred. This is because we have a greater likelihood of obtaining a valid estimate than with **extrapolation**.

TI 84 + $LinReg(ax + b)$	Casio fx9860 series, fx-CG20, fx-CG50 $LinReg(ax + b)$
Press **STAT** to select **EDIT**. **Fill out the two Lists (L_1, L_2)** Press **STAT** to select **CALC.** Press **▼** several times to select $4: LinReg(ax + b)$ Press **ENTER**, then fill out all the required information and press **Calculate.** **Note:** If the correlation coefficient is not displayed then do the following steps: Press **MODE** and then set the option **STATDIAGNOSTICS** to **ON**.	Press **MENU** **STATISTICS** **EXE** **Fill out the two Lists (List 1, List 2)** Press **CALC (F2)** **REG (F3)** **X (F1)** **ax+b (F1)**

Spearman's rank correlation coefficient (r_s)

The **Spearman's rank correlation coefficient (r_s)** measures the relation between two variables when data in the form of rank orders are available. Spearman's rank correlation coefficient is less sensitive to outliers than Pearson's product-moment correlation coefficient because it is based on ranks, and it is sometimes used to evaluate a correlation when outliers are present.

The Spearman's rank correlation coefficient, for example, could be used to determine the extent of agreement between boys and girls concerning their preference ranking of 5 different video games. A correlation coefficient of **1** would indicate **complete agreement**, a coefficient of **−1** would show **complete disagreement**, and a coefficient of **0** would imply that rankings are **unrelated**.

Example 1

	Video game A	Video game B	Video game C	Video game D	Video game E
Boys (x)	3	2	5	1	4
Girls (y)	4	3	5	1	2

The **Spearman rank correlation coefficient (r_s)** for the above bivariate data can be obtained by calculating (<u>only by using a GDC</u>) the **Pearson's product-moment correlation coefficient (r)**.

In this example $r_s = r = 0.7$, which indicates a moderate positive correlation.

Example 2

Consider the following data of 10 students on percentage scores obtained in a Math and Biology examination:

Math (x)	80	75	78	80	65	67	54	79	89	92
Biology (y)	85	77	80	77	70	72	67	86	92	93

We then need to rank the scores for Math and Biology separately. The percentage score with the greatest value should be labeled "1", and the lowest score should be labeled "10". However, we notice that there are two students that obtained the same score (80) in the Math exam, and the same happened in the Biology exam (77). The two values of Math scores would have ranks 3 and 4; hence, both have a joint rank of $\frac{3+4}{2} = 3.5$. Similarly, the two values of Biology scores would have ranks 6 and 7; hence, both have a joint rank of $\frac{6+7}{2} = 6.5$ If we have two equal values in the data, we must take the average number of ranks they would otherwise have occupied.

Math (x)	3.5	7	6	3.5	9	8	10	5	2	1
Biology (y)	4	6.5	5	6.5	9	8	10	3	2	1

The **Spearman rank correlation coefficient** (r_s) in this example is $r_s = 0.912$, which indicates a strong positive correlation.

Notes:

1) **Pearson's product-moment correlation coefficient** evaluates the **linear relationship** between two continuous variables.

2) **Spearman's rank correlation coefficient** evaluates the **monotonic relationship** between two variables. In a monotonic relationship, the variables tend to change together (e.g., when one variable increases, the other one also increases), **but not necessarily in a linear way.**

3) When the relationship between the two variables is **not linear** (visually assessed by a scatter plot), it is more appropriate to use **Spearman's rank correlation coefficient**.

4) Apart from the descending order method described above (i.e., the highest value is labeled by "1", and so on), we can also use the ascending order ranking (i.e., we assign the rank "1" to the lowest value, "2" to the second-lowest, and so on).

Chi-square $\left(\chi^2\right)$ test

The **chi-square $\left(\chi^2\right)$ test** provides a method for testing the association between the row and column variables in a two-way table. This test is based on a test statistic that measures the difference between the observed values we obtained from our sample and the expected values we have calculated.

$$\chi^2_{calc} = \sum \frac{(f_o - f_e)^2}{f_e}$$

Where f_o is the observed frequency and f_e is the expected frequency.

- A chi-square test is designed to analyze categorical data. That means that the data has been counted and divided into categories.

- For a contingency table that has **r rows** and **c columns**, the number of **degrees of freedom (df)**, is given by the following formula:

$$df = (r - 1)(c - 1)$$

- The **significance level** indicates the minimum acceptable probability that the two variables are independent. The significance level defined for a study, α, is the probability of the study rejecting the null hypothesis, given that it was true. It is usually set at **1%, 5%, or 10%.**

- Suppose that **variable A** has r levels, and **variable B** has c levels. The null hypothesis states that knowing the level of **variable A** does not help you predict the level of **variable B**. That is, the variables are **independent.**

  The alternative hypothesis is that knowing the level of variable A can help you predict the level of variable B. That is, the variables are **not independent.**

H_o: Variable A and variable B are independent.

H_a: Variable A and variable B are <u>not</u> independent.

- If $\chi^2_{calc} > \chi^2_{critical}$ then we **reject the null hypothesis H_o** otherwise, we do not reject H_o.

- Another way to reject or not the null hypothesis is by using the **p-value**. If the **p-value is less than the significance level (α)** then we **reject H_o** otherwise, we do not reject H_o.

Note: The chi-square test's two limitations below are not required for the examinations but should only be considered in the internal assessment.

- The chi-square test **is sensitive to small frequencies** in the cells of tables. Generally, when the **expected frequency** in a cell of a table **is less or equal to 5**, the chi-square test can lead to **erroneous conclusions**. This issue can be resolved by combining two or more categories to have expected frequencies more than or equal to **5**.

- If the degree of freedom is **1**, then Yates's continuity correction should be applied.

Example

A hundred secondary school boys and girls are asked which is their favorite color, orange, blue or brown. The results are given in the contingency table below, classified by gender.

	Orange	Blue	Brown	Totals
Boys	10	16	20	46
Girls	27	10	18	55
Totals	37	26	38	101

A χ^2 test, at the **5%** significance level, is performed to decide whether the favorite color is independent of gender.

a) State the null and the alternative hypothesis

b) Show that the expected frequency of a **girl's** favorite color being **orange** is **20.15**.

c) Write down the degrees of freedom.

d) Write down the chi-square value and the p-value for this data.

e) If the critical value is **5.99**, decide whether we reject or not the null H_0 hypothesis.

Solution

a) H_0: Favorite color is independent of gender.

H_1: Favorite color is not independent of gender.

b) The expected frequency of a **girl's** favorite color being **orange** is: $\frac{55}{101} \times \frac{37}{101} \times 101 = 20.15$.

c) $df = (rows - 1)(columns - 1) = (2 - 1)(3 - 1) = 1 \times 2 = 2$.

d) By using a GDC: $\chi^2_{calc} = 8.5667$ and $p - value = 0.013796$

e) Since $8.5667 > 5.99$ or $0.013796 < 0.05$, we **reject** the null (H_0) hypothesis, which means that the favorite color **is not independent** of gender.

The chi-square (χ^2) goodness of fit test

The chi-square (χ^2) goodness of fit test is used to find out how the observed data are significantly different from the expected values. For a χ^2 goodness of fit test, the hypotheses take the following form.

H_o: The data are consistent with a specified distribution.

H_a: The data are not consistent with a specified distribution.

- The **significance level (α)** is usually set at 1%, 5%, or 10%.

- If $\chi^2_{calc} > \chi^2_{critical}$ then we **reject the null hypothesis H_o**.

- Another way to reject or not the null hypothesis is by using the **p-value**. If the **p-value is less** than the **significance level (α)** then we **reject H_o** otherwise, we do not reject H_o.

The number of **degrees of freedom** (df) for a chi-square goodness of fit test is just one less than the number of categories (n) of the examined variable.

$$df = n - 1$$

Example

Suppose we wish to test if a die is fair or not. We roll the die **60** times and get the following "observed" results.

Roll	1	2	3	4	5	6
Observed	8	14	9	8	10	11

a) Write down the table of expected values, given that the die is fair.

b) Write down the number of degrees of freedom.

c) Conduct a χ^2 goodness of fit test at the 5% significance level to find out whether the observed data fit a **uniform distribution**.

d) The critical value for this test is **11.07** . State the conclusion for the test.

Solution

a)

Roll	1	2	3	4	5	6
Observed	8	14	9	8	10	11
Expected	10	10	10	10	10	10

b) $df = n - 1 = 6 - 1 = 5$

c) H_0: The data fits a uniform distribution.

H_1: The data does not fit a uniform distribution.

By using a GDC, we find $\chi^2 = 2.60$ and $p - value = 0.761$

d) Since $2.60 < 11.07$ or $0.761 > 0.05$, we do not **reject** the null (H_0) hypothesis, which means that the data fits a uniform distribution.

TI 84 +		Casio fx9860 series, fx-CG20, fx-CG50	
$X^2 - $ Test	Press $\boxed{\text{STAT}}$ $\boxed{\blacktriangleright}$ $\boxed{\blacktriangleright}$ to select **TESTS.** Press $\boxed{\blacktriangledown}$ several times to select **C:** $X^2 - $ Test or	$X^2 - $ Test	Press $\boxed{\text{MENU}}$ $\boxed{\text{STATISTICS}}$ $\boxed{\text{EXE}}$ $\boxed{\text{TEST (F3)}}$ $\boxed{\text{CHI (F3)}}$ $\boxed{\text{2WAY (F2)}}$ and then fill out all the required information.
$X^2 \text{GOF} - $ Test	**D:** $X^2 \text{GOF} - $ Test Press $\boxed{\text{ENTER}}$ and then fill out all the required information.	$X^2 \text{GOF}$ Test	Press $\boxed{\text{MENU}}$ $\boxed{\text{STATISTICS}}$ $\boxed{\text{EXE}}$ $\boxed{\text{TEST (F3)}}$ $\boxed{\text{CHI (F3)}}$ $\boxed{\text{GOF (F1)}}$ and then fill out all the required information.

t - test

The two-sample t-test is used to determine whether the two sample means $\bar{x}_1, \bar{x}_2$ are equal. It relies on the assumption that the data of each group are sampled from a normal distribution.

The **null hypothesis** $\qquad H_0: \bar{x}_1 = \bar{x}_2$

The **alternative hypothesis** $\quad H_1: \bar{x}_1 \neq \bar{x}_2$ (two-tailed test)

or $\quad H_1: \bar{x}_1 > \bar{x}_2$ (one-tailed test) or $\quad H_1: \bar{x}_1 < \bar{x}_2$ (one-tailed test)

Notes:

1) The two samples will be **unpaired**, and the population variance will be unknown.

The paired t-test compares the same group at two different times, while the unpaired t-test (Student's test) compares two different groups.

2) Apart from the assumption that the distribution of the two groups is **normal**, we also assume that **the two groups have the same variance**.

Example

The students of a math class have been divided into two groups A and B. The students sit the same math test but under different conditions. The students from group A are given instructions before taking the test, while the students from group B are not given instructions. Small samples are then drawn from each group, and their marks are recorded. The marks are normally distributed.

The marks are as follows:

Group A	43	41	48	36	38	33	42	43	37	45
Group B	41	36	35	32	42	48	29	35	33	34

a) State the null and alternative hypotheses.
b) Test at the **5%** significance level, whether or not the mean mark obtained by the students from group A is greater than the mean mark obtained by the students from group B.
   Hint: find the p-value for this t-test.

Solution

a) H_0: $\bar{x}_1 = \bar{x}_2$ (there is no difference between the marks in group A and the marks in group B)

H_1: $\bar{x}_1 > \bar{x}_2$ (the marks obtained by the students from group A are greater than the marks by the students from group B \ one-tailed test)

b) Since the p-value 0.04498 is less than 0.05 (5% significance level), we reject the null (H_0) Hypothesis, which means that the marks obtained by the students from group A (given instructions) are greater than the marks by the students from group B (not given instructions).

The t-value is 1.796. Both values (p-value and t-value) were calculated by only using a GDC.

TI 84 +		Casio fx9860 series, fx-CG20, fx-CG50	
$2 - SampTTest$	Press STAT ► ► to select **TESTS**. Press ▼ several times to select **4: 2-SampTTest..** Press ENTER and then fill out all the required information.	$t -$ Test	Press MENU STATISTICS EXE then fill out the two lists Press TEST (F3) t (F2) 2-SAMPLE (F2) and then fill out all the required information.

Voronoi Diagrams

We often have to locate the closest grocery store, restaurant, or hospital. A map separated into cells, each of which covers the region nearest to a specific center, can help us in our search. This map is known as a diagram named Voronoi after the Russian mathematician Georgy Voronoi (1868-1908).

A Voronoi diagram is a mathematical tool applied across many scientific disciplines. These include medical diagnosis, astrophysics, network analysis, computer graphics, and artificial intelligence. In brief, Voronoi diagrams are proximity diagrams that divide a plane into areas based on the distance to points in the particular part of that plane. Given a set of **sites** in a space, a Voronoi diagram, as shown in the following figure, partitions that space in **cells** — one cell for each site. Each cell contains all the points that are closer to that site than to any other. The Voronoi diagram is the subdivision of the plane into cells, one for each site. A Voronoi diagram tessellates; that is, it consists of polygons that fill an entire space without gaps or overlaps. A line is called an edge between two areas. **Edges** lie along perpendicular bisectors of pairs of point sites. An edge is always divided into two areas precisely. Where three or more borders meet, a **vertex** is called. At least three edges at the vertex are always there, but there can be more. The line around each site includes the region of the site. The sites are always convex sets (A convex set is a set of points such that, given any two points in that set, the line segment joining them lies entirely within that set).

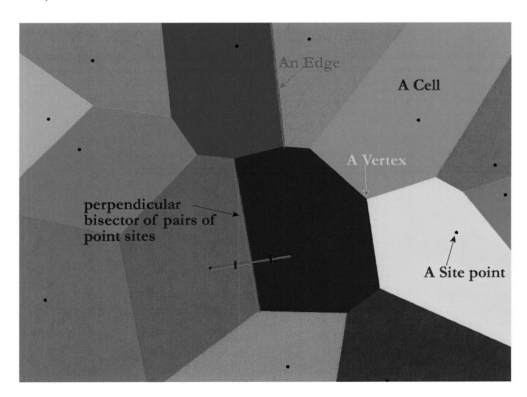